WEMYSS TRAMS AND EARLY BUSES

ALAN BROTCHIE

Stenlake Publishing Ltd.

© 2023 Alan Brotchie
First Published in the United Kingdom, 2023
Stenlake Publishing Limited
54-58 Mill Square, Catrine, KA5 6RD
www.stenlake.co.uk

ISBN 978-1-84033-922-2

The publishers regret that they cannot supply copies of any pictures featured in this book.

Printed by
P2D Books, 1 Newlands Rd,
Westoning, Bedford MK45 5LD

Wemyss car 9 is about to turn right from St. Clair Street into Junction Road, Kirkcaldy, in this 1907 scene. The rails of Kirkcaldy's original tram route continue ahead but were not included in the through operation agreement. The Corporation's 'Lower Route' cars to Linktown terminus normally ran as far as at the loop seen in the distance.

Acknowledgements

Particularly to Mr Michael Wemyss, and Mr Charles Wemyss for access to contemporary estate records, aided by Miss Elizabeth Wemyss and Miss Denise Lyon at the Red House. To Andrew Dowsey and his assistants at Fife Archives in Glenrothes; Eric Eunson and Bill Fiet for use of their photographic collections and David Heathcote for permission to use images recorded by his father, George Heathcote.

Note: Two photographic commissions were made of the newly completed project, one set by the contractor, Bruce Peebles taken 4 September 1906, the other by John R Patrick of Leven (an outstanding photographer established 1866) these probably for Stephen Sellon and taken a few days later. The first set was printed for me from the original negatives; Patrick's work is reproduced as originally made, in sepia tones. They provide an unrivalled record of the area at that time.

Preface

When Captain Randolph Gordon Erskine Wemyss of Wemyss was laird of Wemyss Castle in Fife his insight, instigation, promotion, and ultimately his finance saw his concept of a local inter-urban style tramway carried from a dream to a reality. The funding of a project of this nature by just one remarkable individual was probably unique in Great Britain. However the Wemyss tramway was exceptional in more ways than just this one. His determination to ensure beneficial development of his mineral-rich inheritance followed a period of fifteen years while it had been maintained by trustees until he came of age on 11th July 1879. Thereafter he rapidly pursued a much more positive programme of exploitation of his underground minerals. Connection of the estate and coastal villages to the North British Railway (NBR) at Thornton had been initiated by his father in the final year of his life (with initial plans prepared in 1865), but it was Randolph through his energetic factor, Joseph Budge, who saw this project to a successful conclusion. The line opened as the Wemyss and Buckhaven Railway with fit and proper ceremony on 1st August 1881, but unfortunately the seeds were soon sown for a prolonged time of frustration in dealing with the NBR's management and board of directors.

My long-standing interest in the development of passenger transport in this part of Fife gave me the chance to meet many individuals who had given service on the tramway, local buses and the estate railway. The memories of these participants added flesh to my 1976 account of the tram undertaking; with a history of the estate railway following in 1998. It is now possible to add to the tram story, with detail from newspapers now available online. Minute books and contemporary letter books are now available which were not in the public domain then. When undertaking these works there was no online facility, but I was permitted by editors of local journals to examine 'hard copies' of their newsprint. It is also possible now to improve photographic coverage beyond the limitations of the previous works.

Positive assistance then from the Wemyss Development Company's officers allowed investigation into their railway. This undertaking – despite considerable establishment resistance – but with the absolute determination of the laird, in 1947 retained its separate existence outwith the nationalised NCB coal industry. Only with encouragement from Captain Michael Wemyss, Mr Michael J Wemyss and the company's dedicated managers – in particular Mr Andrew Walker – was a detailed history of the Wemyss Private Railway possible. This was made more comprehensive with their active participation.

For this revisiting of the tram story – another major factor of the improvement of the estate – Mr Michael and Mr Charles Wemyss have again greatly assisted my inquiries. At the estate office, Elizabeth Wemyss and Denise Lyon have always been most welcoming. Specialised assistance on the omnibus scene has come from Richard Gadsby while Eric Eunson and Bill Fiet have given unrestricted access to their local history collections. Many old postcards are reproduced in colour as they were originally, and several other photographs have been digitally enhanced in colour by Dr Mike Mitchell. Thanks also to David Heathcote and Alan Simpson. The Omnibus Society have recently (June 2020) published their comprehensive study of buses in the former Fife area. Technical detail given there is much more than here, listing every known operator and every known vehicle in the area. I have where possible included information which has surfaced in the last four decades, without more repetition than is necessary to complete the picture.

"The Laird". Randolph Gordon Erskine Wemyss (11th July 1858 – 17th July 1908). Original painting by an unknown artist owned privately, this image by Fiona Hutchings.

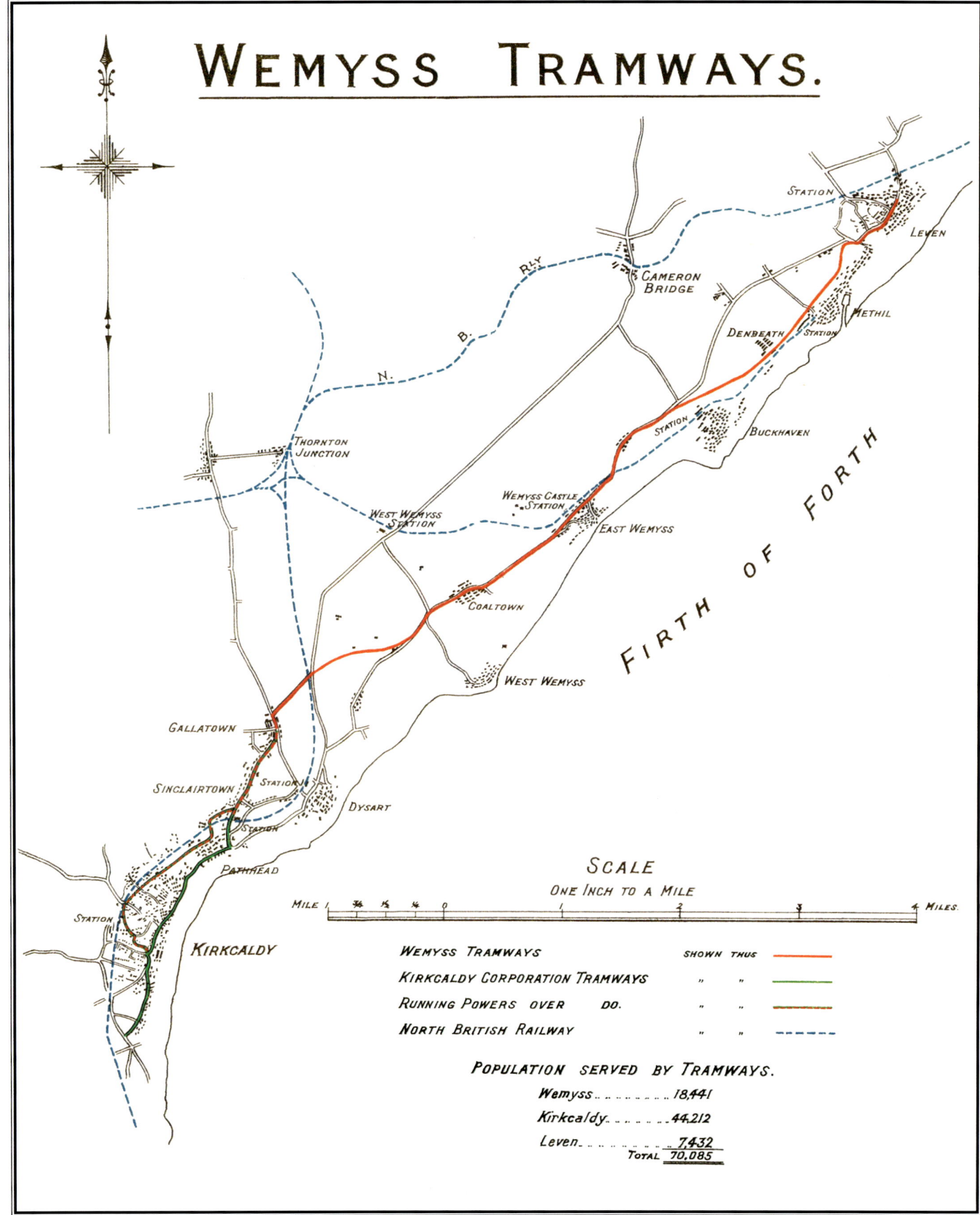

The route plan which accompanied the initial prospectus.

Introduction

At the end of the 19th century it appeared that an answer had been found to new problem associated with movement of the large labour forces required to serve major industries. Electric tramways had progressed beyond the experimental stage and had become a reliable means to allow the workforce to live at a distance from their place of work. Now the tram became cheap enough to act as the 'working man's carriage'. In the preceding century fares were mostly beyond the reach of workers for regular and frequent travel. With a practicable and financially justifiable method of electric propulsion, using an overhead wire for power distribution to the vehicle, there followed an electric tram construction boom. This attracted interest, initially by the Corporation of Kirkcaldy, in Fife. Well known as the 'Lang Toun', at the end of the 19th century the town straggled as a continuous ribbon development some three miles from Linktown at its southern extremity to Gallatown in the north. It was said there were then Gallatown residents who had never visited Kirkcaldy's High Street; they had no need to, why should they? This changed when their council decided in the final year of the 19th century to build an electric tramway connecting the extremities of the burgh; this opened for business on 28th February 1903.

Randolph Erskine Wemyss became laird of the 6,100 acre Wemyss estates in Fife at the age of five, in 1863, but these were managed for the next sixteen years until his coming of age by trustees, led by his mother. The young visionary laird proved to be determined to develop the mineral wealth beneath his estate. A relatively limited area of this had been leased by the trustees who, in 1870 appointed as the estate factor Joseph Budge from the Altyre Estate near Forres in Morayshire. Later charged as Mr Wemyss' Commissioner, Budge was to take a pivotal role in day to day operation in many, if not most, of the Wemyss enterprises. He was frequently given a mere broad brush outline of the laird's wishes, and it was he who had the responsibility and confidence to achieve success. With regard to the creation of the Wemyss and Buckhaven railway, he purchased the entire permanent way materials for issue to the contractor, plus similar subsequent arrangements for other connecting lines and sidings. From 1881 general management of the railways plus docks at Methil and Leven was devolved to Budge. He similarly played a fundamental role in development and execution of the Laird's tramway schemes. Budge served the estate for 37 years, with oversight into all of Mr Wemyss' many developmental proposals. In addition to these demands on his time, he served on the boards of directors of the Coal Company, the Tramway Company and many local charitable organisations, plus as a Justice of the Peace. His interpretation of the laird's vision created the tramway, which played a major part in the development of the communities it served. Many of the documents which remain have provided fascinating detail, incorporated here to add flesh to earlier accounts.

"*The Factor*". Joseph Budge; (d. 20th October 1914). Initially employed as factor, he later became Mr Wemyss' Commissioner.

Car number 1 heading east down Leven High Street on the first day of public operation in 1906. On the right, of the pair, the young man on the left is Jim Bryant, a lifelong enthusiast of public transport and who collected scenes of 'his' trams avidly, and whose collection [and enthusiasm] was passed to the author. This book – the second edition and much revised – history of the line is dedicated to his memory.

Although printed in the *Dundee Courier & Advertiser* on Saturday 30 January 1932, indicating that the Wemyss trams were to cease operation "this weekend", the timing of the event was supposed to be a well kept secret! The view of 'English' bogie car 20 outside Aberhill Depot was in fact probably taken on the previous day.

Tramway History

Kirkcaldy's municipally sponsored electric tramway was conceived in the latter years of the 19th century and came to fruition on 28th February 1902. The 3-mile long route with ten conventional double-deck vehicles provided a guaranteed demand for the newly built municipal power station on Victoria Road. Its success soon attracted the attention of Mr Wemyss, with the earliest record of his interest noted just six months later, in August of that same year. He quickly grasped the benefits to his other enterprises from this ability to tap into an enhanced area for his workforce. Added to this was the considerable additional benefit of a rapid connection for passengers to travel by way of The NBR station at Kirkcaldy. Considerable pressure was applied by the council of the then separate Burgh of Dysart for this new connection to run through their burgh, but with a stated purpose of improving communication with Kirkcaldy main line railway, this was resisted.

Mr Wemyss' initial thoughts were to have this tramway built by the NBR, but they gave the idea short shrift, viewing it as competition for their passengers using the existing rail branch to Methil. This required a change at Thornton Junction, frequently involving lengthy delay and was most unpopular with local travellers. The railway company also held the opinion that their opposition in Parliament would be sufficient to ensure its failure to gain approval. Undeterred, Wemyss then approached Kirkcaldy Council with a suggestion that if he built the line, they could operate it. At that time they were heavily into extension of their undertaking by building a second line, the so-called 'Upper Route' and declined to be involved beyond the direct needs of the burgh. Following this second rebuff, Mr Wemyss decided that the line should be built, financed initially entirely by himself. Kirkcaldy's second route ran past the town's main rail station and at an early date agreement was reached to allow the Wemyss trams to run into town to terminate at Whytescauseway, only by this line passing this station, not on the town's other lines. Considering Mr Wemyss position in the community, few difficulties were raised by local authorities; major objections coming only from the NBR. This became typical of that company's

Cap badge featuring the family crest, a white swan, and motto *Je Pense*. The device was used on the cars and buses, company seal and for director's passes (in gold).

dealings with Mr Wemyss. When the Wemyss & Buckhaven Railway and Methil Dock were absorbed by the NBR Mr Wemyss was offered (and accepted) a seat on the board. He rapidly became a thorn in the flesh there, particularly regarding the manner in which the railway company planned to advance its interests in East Fife, with regard to coal export and harbour development. He served on the board for ten tumultuous years from early 1889.

Initial surveys for the tramway were carried out in early April 1904 by Thomas Meik & Son, civil engineers from Edinburgh, while Kennedy & Jenkins, the Corporation's consulting electrical engineers, gave advice on power supply issues. Plans were prepared by Meik showing the proposed route of the tramways, required for submission for a Provisional Order, part of the legislative process required by Parliament to authorise new passenger-carrying tramways. Agreement was reached with all the local authorities involved, not difficult when the instigator was the feudal owner of virtually the entire area involved. Track was to be to the least expensive specification possible – which very soon proved to be a false economy involving costly reparative work. Such economies, in an area highly susceptible to mining subsidence, were not a clever idea.

Strong objection to the proposals was made by the NBR, which required a Public Inquiry for the issues to be debated, incurring considerable legal expense, beginning at the end of March 1905. Before this commenced Mr Wemyss replaced Meik as his engineer. He was considered too close to the NBR, being replaced by Stephen Sellon. There can be little doubt that, with regard to expertise in tramway promotion, there were few engineers with Sellon's impressive track record. During the Inquiry he acknowledged that he had been responsible for no less than 500 miles of tramway construction in the previous twelve years.

Stephen Prescot White D'Alté Sellon, an Englishman who early in his career joined the Thomson-Houston Company in the United States, was probably then the most suitable individual to further Mr Wemyss' tramway ambitions. An early associate of Emile Garcke, the founding entrepreneur of the British Electric Traction Company [BET] Sellon devoted most of his career to tramway development, becoming the chief engineer of the BET, then in 1886 establishing his consulting engineering practice in Westminster. One of his early projects was the Wolverton & Stony Stratford Tramway in Buckinghamshire which he described as "... the first tramway system in this country designed for comprehensively dealing with goods traffic." He travelled extensively in Europe and the USA studying early electric traction developments and installations and devoted his career to the furtherance of electric tramways in the United Kingdom. In this role he was responsible for the first such installation in the country using overhead wire power supply, at Roundhay in Leeds. From there he continued to Coventry then Dover. His evidence was crucial in

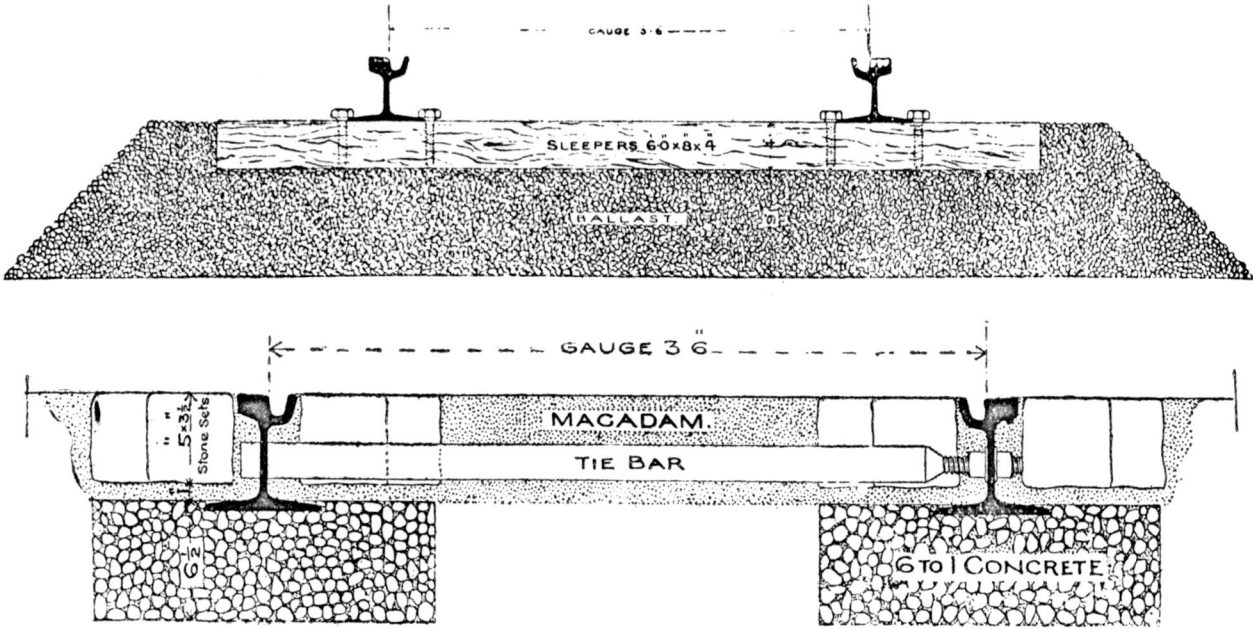

Contrasting details of track construction. Above is the sleeper construction for off-road and light railway track, using colliery ash for ballast. Below, details for tracks laid in public roads, showing the unreinforced mass concrete strip foundations. This was not a suitable form of construction for ground highly susceptible to mining subsidence.

achieving the Light Railways Act of 1896 under which several tramways were promoted that were unlikely to have met the criteria imposed by existing legislation, the 1870 Tramways Act, which was drafted before dawning of the electric era. Sellon was subsequently engineer for the South Staffordshire and the Potteries tramways. His employment by Mr Wemyss soon saw him appointed managing director of the Wemyss tramways with an emolument of £500 per annum, but from this he had to provide a superintendent of operations.

The main thrust of the NBR's objections lay in the competition they perceived for goods traffic along the proposed line through the Wemyss estates. This concern might have been confirmed by the evidence of Captain Gordon, manager of the Wemyss Coal Company, who advised the Inquiry of the benefits and savings which would be achieved in using the line for distribution of free coal to their employees. This not inconsiderable traffic for a workforce of – even prior to the major expansion schemes which Wemyss had in prospect – over 1,000 men underground and on the surface. Each employee was entitled to an allowance of free coal (the sole form of heating and occasional hot water) of 8 cwt per fortnight, or 12 cwt if there were two employees in a household. Distribution of this had to be paid for: the coal might be free, but delivery was not. By horse and cart this cost 2s 3d per load but by transferring this operation to the tramway Gordon estimated an "enormous convenience". Mr Wemyss later also envisaged distribution of milk from his reconstructed 'model' farm at Bowhouse, but neither of these ideas became reality. The Inquiry found in favour of the tramway proposal, but in order to protect the interests of the railway company, a limit was imposed on the maximum weight to be carried as 'goods' of 500 lbs, or less than ¼ ton, thus precluding any serious idea of cartage of coal by tram. An objection by the water authorities was dismissed. They had expressed concern regarding potential for electrolytic damage to their pipes by stray electrical current, but were advised that adequate regulation already existed to cover this matter.

The Wemyss and District Tramways Act became law, granted the Royal Assent on 11th August 1905, giving formal permission for construction of just under eight miles of tramways and tramroads (tramways not on public roads). Track was to be generally single, with passing places about every half mile, another saving which was soon proved to be a false economy. Included in these authorised lines was a branch, ¾ mile long, down the Boreland Road to form a connection with any extension of Kirkcaldy's tramways to Dysart. The council of this small burgh was very anxious to join in to the benefits anticipated, but Mr Wemyss (and Kirkcaldy Corporation) were less so. As it happened this Wemyss branch was never built, but Dysart did get a tram connection to Kirkcaldy five years later.

Before work had even started, in April 1905 Mr. Wemyss had Stephen Sellon report on a network of passenger tramways to join together other populous and coal mining areas of West Fife. In 1903 the Royal Navy started planning at Rosyth of an entirely new east coast naval base and in anticipation of the development which would undoubtedly follow, a tramway to link existing centres of population was another of Mr Wemyss' intentions. From Rosyth the 23-mile 'main line' was to go by Dunfermline, Cowdenbeath, Lochgelly, Auchterderran, Kinglassie, Bowhill, Cardenden and Thornton and on to meet Kirkcaldy's tramways at Gallatown. Branches to link to Lochore, Kelty, Cardenden and probably also Markinch were planned. Construction of the Naval base commenced in 1909. Another proposal at this time was for a tram line from Rosyth by the coastal road via Inverkeithing and Burntisland to join the westernmost terminus of the Kirkcaldy lines. This never materialised, but a large proportion of the former proposal was built (by Balfour Beatty Limited and with Sellon as engineer) between 1909 and 1918. Intervention of the First World War and motor bus proliferation thereafter ended any possibility that the additional lines might be built. Parliamentary Authority for construction was thereafter renewed on a regular basis.

Mr. Sellon's local resident engineer was an experienced civil engineer, James McDonald Angus, who had worldwide career in tramway contracts. Since 1897 he had been employed on a succession of BET tramway projects. He and William Griffiths of London, appointed as contractor for the track laying, had worked together previously, including on the Potteries tramways in Staffordshire (where the cars provided bore a strong similarity to those in hand for the Wemyss line). By May of 1905 engineers were setting out the line of the new 'tramway' road to replace the Low Road along the shore, where Mr Wemyss was planning a large acreage of railway sidings to serve the development of his new Wellesley Colliery. It was agreed that to facilitate feuing and the layout of subsequent new roads, the new road '... would be set out parallel to the existing Bowling Club at Methil ...'. This, to accommodate the positions of two new mineral railway bridges required that Bayview Park, home to East Fife Football Club since formation in 1903, had to be moved several yards to the north!

Without waiting for the formality of Royal Assent to the Act, on 3rd July Griffiths, the contractor appointed by Sellon, started work on the largest piece of engineering. A curved cutting almost twenty feet deep at a gradient of 1:22 was needed down through Kinarchie Braes from Aberhill to meet Methilhaven

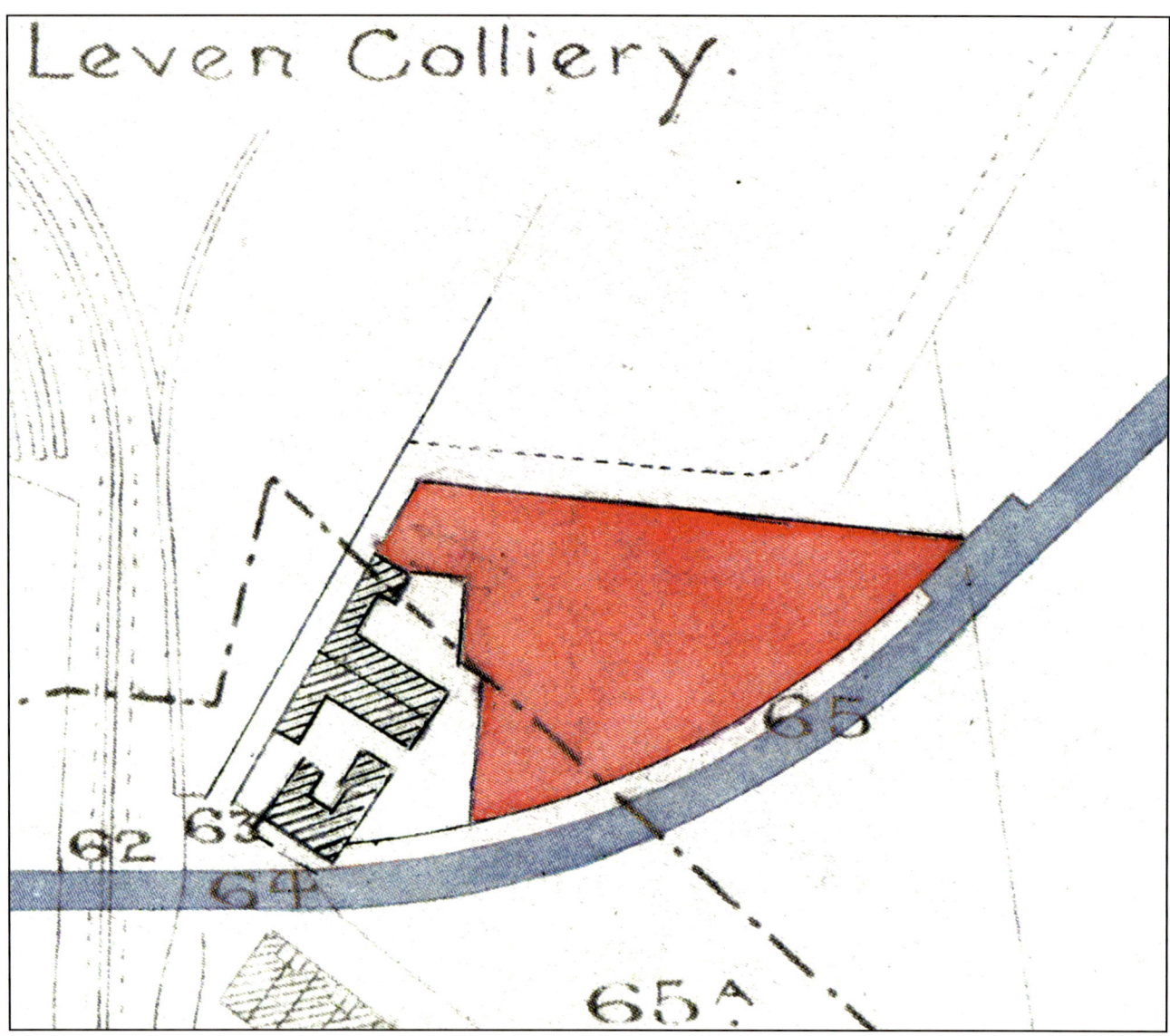

Detail of ground acquisition at Aberhill Farm for the new road, tramway and depot. The existing buildings were not at this time purchased, but were soon obtained for use as offices etc.

Road at the end of the 'Bawbee Brig' over the River Leven. This bridge was always known thus from the half penny toll extracted for foot travellers to cross – until this levy was removed by statute in 1870. Griffiths, who was working on a cost plus basis, quickly put a large squad of navvies to work and excavation was completed before the end of August. The two rail bridges necessary for the new tram route were also built within the same remarkably compressed timescale.

During early September the first cargoes of tramway rails and sleepers arrived at Methil Docks and were rapidly moved to be stockpiled along the tram route. The off-road lengths of track, of sleeper construction, were laid on ballast consisting of ash from boilers at the various collieries. As the track was completed, Griffiths requested, and was granted, permission to use these new lines for wagons to distribute this material. One of his tipping wagons was retained for subsequent permanent way maintenance.

In the original proposals the intention was apparent that the tram depot and power station would be in East Wemyss, on the north side of the main road adjacent to the access road to Wemyss Castle Station. This was changed, for whatever reason is unclear, so that the car depot should be built adjoining Aberhill Farm, alongside the Fife Coal Company Leven Colliery (No. 1 and No. 2 Pits) near Methil Brickworks. The unassuming depot structure with three tracks (but no maintenance facilities whatsoever) was built rapidly by Thomas Topping Ltd of Edinburgh. Instructions were issued to the builder to "... build the car shed to the east of the farm and leave the dairyman severely alone ...". This remained a working farm, but before long the fields were taken over for housing and the remaining farm buildings were adapted for use as the tramway company offices and stores. It had been intended that all maintenance for the Wemyss' cars would be undertaken at the workshops of Kirkcaldy Corporation at Gallatown Depot. Topping also built the generating station at Denbeath, just west of the new Baum coal washing plant, the largest such facility in the world when constructed.

The Coal Company was to own the power plant and supply the tramway at the advantageous rate of 1¼d per BThU, reducing to 1d when demand exceeded 100,000 units per annum. The contract for all electrical equipment, including cars, was placed with Bruce Peebles & Co of Edinburgh. Generating gear supplied by them consisted of two Bellis & Morcom engines each coupled to a BT-H generator; steam supplied from the colliery's boilers. Up to this stage the entire enterprise had been funded directly by Mr Wemyss, referred to as 'The Contractor'. This role had to be renounced, however, as the Board of Trade, as a condition of their approval, required that the undertaking was transferred to a standalone company. Good progress was made initially and it was hoped that the line could open by Christmas. This was wishful thinking since bad weather caused delay.

The original agreement between Randolph Wemyss and Bruce Peebles for the electrical equipment of the line calls *inter alia* for six cars plus three bogie cars, this latter item deleted by a pen stroke through. Operation of these vehicles, fundamentally for miners – was a major consideration from the first, and they were specified to have ' ... open sides, waterproof curtains with sixteen cross-bench seats for four and vestibuled driving platforms'. This does give a good description of these cars as built, although the fact that they appear to have had a section eighteen inches long cut off one end of the saloon is strange; they could easily have been built with all the saloon sections of the same width.

A high degree of optimism was in evidence at the first meeting of the board of directors of the new company held on 9th November following, when it was stated that the line would open for traffic on or before 31st March 1906 – but even this proved to be

Tramway power plant newly-constructed at Wellesley Colliery. The two British Thomson-Houston generators are driven by Bellis & Morcom engines with output at 500 – 550 volts.

over ambitious. That initial meeting laid the ground for much of which was to follow. Stephen Sellon's engagement as engineer was transferred to the new company. In addition he also as manager had the responsibility for providing a competent superintendent. This role went to James Angus. Unfortunately this highly-popular individual was in post for just two months, as he died during the following January. His successor, Lionel C Knocker was less admired, quickly establishing a reputation for 'rubbing people up the wrong way'.

The board confirmed that the company's capital would be £30,000 in 4½% mortgage debenture stock, 6,000 6% preference shares of £5 each, plus 10,000 ordinary shares of value £1. These (but not including the ordinary shares) were offered for subscription on 13th November, with a closing date just three days later – "... or before". The expectation of a stampede of enthusiastic investors failed to materialise, and by the closing time, only 3,270 debentures and 1,129 preference shares were allotted, most locally. The largest tranche of these was £1,000 of debentures taken by the Fife Coal Company. Mr Wemyss' earlier exhortation to "... risk a sporting sovereign..." seems to have fallen on deaf ears. The 'consideration' to the Contractor for building the line was to be £65,000 (all of the Ordinary Shares plus £55,000 in cash). Any balance not publicly taken up was to be allocated to the contractor in lieu of the cash sum and Mr Wemyss was consequently allocated £16,000 of debenture stock.

Construction of the tramway, which had started with such gusto, slowed and projected completion dates – always ambitious – had to be pushed back. Mr Wemyss had set a second (or third) date of the July holiday period, but even that slipped, due it was said to Bruce Peebles delay in completing overhead wiring. Feeders for power distribution to the overhead were also slung above ground as another cost saving. Eventually on 28th July one of the newly-delivered cars was hauled by three farm horses along the line to Gallatown "... to test the gauge ..." was the explanation. The first trial under power took place ten days later, attracting the admiration of large numbers of wondering spectators. This initial car was driven by James Fisher, superintendent of the Kirkcaldy tramways, indicating the close harmony already existing between the two undertakings. Fisher was a brother of Peter Fisher, manager of Dundee's tramways. Several meetings had been held to ensure that the new Wemyss cars were compatible with those already in use on the Kirkcaldy lines, so that there would be no hitch when through operation into the 'Lang Toun' commenced. Also at this time of rapid recruitment of operating staff several significant senior appointments were made. In July William Dawson was appointed chief cashier, later accountant, a position held until 1911 when he was promoted to became the general manager; a role held for the remaining existence of the tramway. The role of traffic superintendent went to William Mittell who was to lose his life in Africa during the First World War. His son continued the family transport tradition, appointed chief draughtsman to Edinburgh's tramways from 1920, with thereafter a major input into the electrification of that system and the continuous development of their rolling stock culminating in the (then) highly experimental tram number 180 which incorporated much ground-breaking use of aluminium in its bodywork.

The next milestone was the statuary inspection by the Board of Trade which unusually was split into two visits. The first, by Mr Trotter dealt with the electrical elements and was held on 10th August, passing the line as satisfactory. He was followed four days later by Colonel von Donop who examined permanent way and rolling stock. All was well except that the cars had not yet been fitted with their lifeguards – these still en route from the makers. Permission to open was delayed until this work was complete. On much of the line a maximum speed of 15 mph was authorised, faster by some 20% than usually permitted for 'town' tracks. He also insisted that a full-time gate keeper be provided at each mineral railway level crossing.

With little advance warning and completely without ceremony, public service commenced at 6.40 am on Saturday 25th August, five cars providing a nominal 20 minute frequency service as far as the Gallatown. The previous day, in a foretaste of things to come, one of the Wemyss cars was given a trial run along the full length of the new line, again driven by Mr Fisher, but then continued along the Corporation lines as far as Whytescauseway. Being the weekend, there was no workers' service, but this gave full opportunity for the new spectacle to be enjoyed to the full – and the novelty was taken full advantage of. The following day it was found that over 4000 tickets had been issued, with the cars – and their crews – almost overwhelmed by the numbers of joy riders. It was rapidly realised that the length of the line allowed travellers on the Sabbath to qualify as 'bona-fide travellers' and hence entitled to alcoholic refreshment. Publicans were also quick to spot a potential windfall and the price of a quarter gill of *usquebagh* in Leven (on a Sunday) immediately rose from 3d to 6d, a severe deterrent, but not enough to become a total preventative. Exuberance of travellers led to suggestions that trams should not run on Sundays. Therefore to prevent disturbance to church services cars terminated at Mitchell Street, 400 yards short of the terminus also most conveniently immediately beside the Caledonian Hotel.

The following week saw the hectic patronage of the new trams continue unabated and it was rapidly

Driver's power controller as manufactured by Bruce Peebles of Edinburgh for the first batch of cars. Of their flameproof 'PPP' type, it was of conventional design; seen with the front cover open. The controller handle at the top moved the contacts, arranged vertically on the left which regulated the current reaching the motors. The smaller handle to the right determined the direction of travel, forward or reverse. Both handles were removable, and taken to the other end of the car to move in the opposite direction.

Driver's platform of one of the first cars showing the basic equipment. The controller is operated by his left hand, and again, the cover is open. To the right the two vertical hand wheels are brakes, a track brake and a wheel brake. There are expanding scissor gates on both sides of the platform – source of many pinched fingers.

realised that, if this was going to be sustained, more rolling stock would be required. Sellon was instructed during the first week of September to negotiate for four additional cars and two parcel vans. An order was rapidly placed with Brush Ltd for more cars almost identical to the first nine, and Sellon entered into negotiation with the Potteries tramways for purchase of two parcels vans which they had available.

A sumptuous celebratory opening lunch, hosted by Mr Wemyss and his wife Lady Eva, was held for Saturday 8th September, with over 200 of the local great and good (and others) assembled in the banqueting hall of Wemyss Castle. Also present were representatives of those who had built the line, the contractors and the engineers. During the morning, five special cars treated the guests to a trip along the full length of the line, with guided tours laid on of the power station and the depot. In his speech accompanying his toast to the success of the new line, Mr Wemyss stated inter alia "... if they [the NBR] were to build the tramway he would give them the land for nothing. They refused so there was nothing for it but to do it themselves. To his mind there was nothing so important or educative as to give facilities for an ever-increasing population than to move about cheaply. A working man might live in a very nice house, but he could not afford a carriage and pair or a motor car, and the railway communication was such up at Thornton that a man leaving Buckhaven early had to lunch and probably dine at Thornton. When they decided to have their tramways they had a good deal of croakers to contend against. ... When the Bill came up the manager of the Railway Company in the witness-box said the district was so efficiently served that there would be no traffic for the cars. Well, he made a mistake, because in the fortnight the cars had been running they had carried 53,000 people and he thought it would go on increasing. ... in regard to what he hoped would happen in the future, when there would be a system of tramways working through Fife and connecting the various industrial centres, and that when these halcyon days came the tramway would be worked alongside the railway, and feed and help it instead of being a competitor. ... if they had had to carry out the demands the Local Authorities made ... it would have cost another £2000 to £3000 a mile. They had got that cost off, and they got the tramway made at a reasonable rate. It was due to the initiation of Kirkcaldy in building their tramways that enabled them to put one up to join with their end. Kirkcaldy Corporation met

Crowds in Coaltown on the occasion of the official opening on Saturday 8th September 1906. The cars had been in operation for two weeks, since 25th August, but this scene (probably recorded by pioneer photographer John Terras of Markinch) shows the invited guests and others by an entrance to the Wemyss Castle policies.

Probably taken on the same occasion, this card was sent from Coaltown to patient Willie Adamson in Edinburgh Royal Infirmary "... I see John and Peter Birell on the other side but not you ...". These may be the two lads extreme left of the scene.

them in a reasonable, he might say generous, spirit. The result had been that the Kirkcaldy receipts had gone up." Mr Sellon on behalf of the directors then presented Lady Eva with a miniature tramcar controller handle, to which was attached a gold free pass. This, he said, was the only one they intended to issue, while expressing the hope that she might occasionally make use of the tramway cars forsaking her husband's magnificent motor cars. After more toasts, the company dispersed.

Two major problems were immediately apparent; an inadequate number of cars and the difficulties inherent in the single line track as built. The first was addressed by an immediate order to Brush, less than two weeks after the opening, for four more cars, with a short-term solution achieved by hiring from Kirkcaldy Corporation. They had more vehicles than required for their needs and were more than happy to oblige. For 10/- per day each, three Kirkcaldy cars were transferred to the Wemyss lines from 29th September. Trolley heads of the Wemyss pattern were fitted temporarily to these cars, but it was rapidly discovered that the trolley ropes of the single deck cars fouled the top decks of the double deck Kirkcaldy vehicles. To overcome this the Wemyss' cars had their trolley bases raised 2 ft 3 inches. To satisfy the Board of Trade regulations, no passengers were permitted to travel on the upper deck seats of any Kirkcaldy tramway vehicle. The Kirkcaldy cars came with their drivers at 5d per hour but conductors were supplied by the company.

In an attempt to improve – or rather to introduce, – some degree of regularity to operations, a contract was placed for automatic electric colour light signals to control access to single track sections. Supplied by Brecknell, Munro and Rodgers of Bristol, these came into use in March 1907, with immediate operational improvement, meetings on single track sections virtually eliminated. To assist this a new loop was added at Methil Brae crossroads.

Transport of miners to and from their workplace was a prime purpose of the new tramway, and this started (at 4.53 am) on Monday 24th September with cars 'for this special class of worker', plus reserved cars at shift change over times. For this it was necessary to use the brand new cars, but with canvas seat covers in an attempt to keep the wooden seats clean. Pit head baths were an unheard of luxury and men travelled in their working clothes, which, after a day's toil would often be ingrained with coal dust. Cars solely for this traffic had been intended from the start with discussion with the Board of Trade regarding use of bogie trailer cars. The Board had severe objections to this on principle, seeing them as inherently dangerous. In surviving papers this attitude is very apparent, as is Sellon's equal determination to get permission for their use, as he had on the tramways he designed for the Potteries district. The first specification of January 1906 included three open-sided trailer bogie cars for workers' traffic. They were to seat 64, on sixteen cross seats for four passengers; dimensions were given as 40 ft 6 ins over fenders by 5 ft 8 ins wide. However this provision was removed from the tender documents with the number of small cars altered in pen from six to nine. Mr Wemyss wrote to Sellon " ... I would prefer if the Colliery found them ..." – perhaps an odd decision, to lumber the Coal Company with a task clearly outwith their normal business of raising and selling coal. His motivation is unclear, unless it was simply to ensure the financing of these cars was distinct from tramway operation. Existing Coal Company letter books have nothing relating to the purchase of these four cars.

Later that same week (on Thursday 27th) Wemyss cars commenced operation via Kirkcaldy's 'Upper Route' (Victoria Road) to Whytescauseway as contained in the earlier agreement. What could, and perhaps should, have been included, was that this was to be coordinated. However, it was not planned properly in advance, with the result that the Wemyss cars often duplicated, instead of complementing, the Kirkcaldy service. Also track in Rosslyn Street was single, leading to disruption. At 6d for the through journey of 10 miles the new service was immediately popular and it was recorded that Kirkcaldy's receipts rose by £50 per week; a much-needed 'shot in the arm' for the financially struggling Corporation system. The Wemyss cars operated on 15 minute frequency with alternate cars running through to Whytescauseway.

An agreement was also reached to house one or two of the Company cars in the Corporation's Gallatown depot and to operate the early morning services to Leven from here, saving dead mileage. Mr Wemyss at this time proposed to develop Bowhouse Farm to supply milk on a large scale. Facilities were extended and (it was stated) a tram siding and loading bank were built for distribution by the tramway. There is no other evidence that this was completed, or used. The new route from North Lodge across fields to Bowhouse formed a handy shortcut for pedestrians, but since it was unlit, was a highly dangerous habit, leading to several accidents. Mr Wemyss had boards erected at both ends of the private track prohibiting trespass, but with little effect – the short cut was much too convenient.

In the first week of September 1906 Mr Sellon was instructed to obtain parcel vans for the Wemyss line. As engineer to the Potteries tramways he would be well aware that they just happened to have two surplus

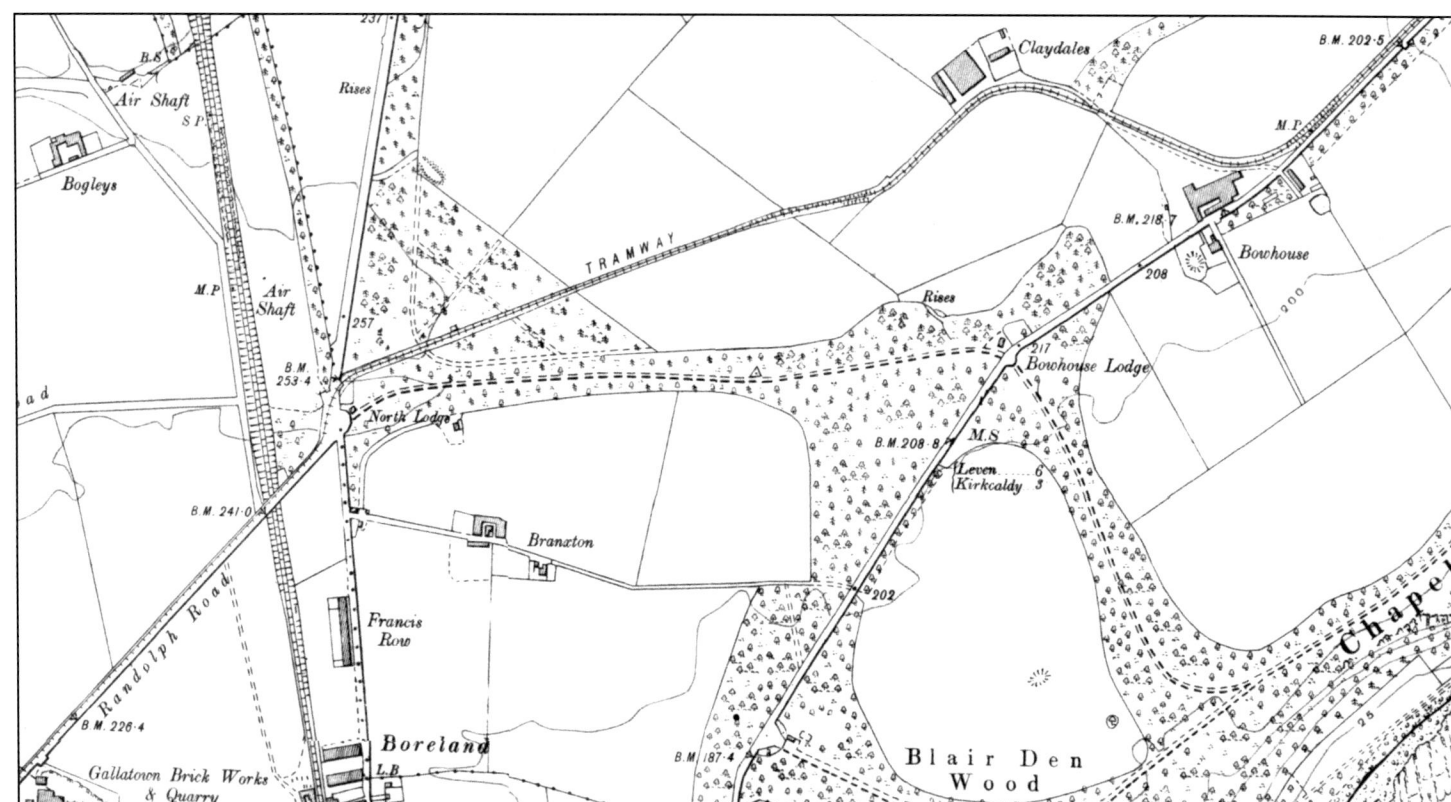

Plan of western end of the tramway, cross country on its private track from North Lodge (of Wemyss Castle) to Bowhouse Farm – the home farm of the estate. From here it continued on private right of way to the north west side of the public road.

vehicles of this nature. Previously it had been assumed, based on circumstantial evidence, that these were the two vans obtained for Wemyss, but this has now been confirmed by entries in the existing cash book. What is not related is that these were of 4 ft 0 ins gauge, and would require to be altered to suit the Wemyss gauge, 6 inches narrower. The pair arrived at the end of January 1907, but were not put to use until 1st May. They cost £142 for the pair, against Sellon's estimate of £170; the NBR paid transport costs of £8.4.10d. Presumably when the costs of regauging and repainting were added in, Sellon was not far off the mark. The Potteries undertaking had ordered six of these little vehicles, probably from the Brush Company, in 1902 but soon advertised two for sale at £100 each (their numbers 113 and 114).

Next to arrive were the Coal Company's cars for workmen's traffic which were delivered in mid-April 1907. It is possible that they were firstly numbered 1 to 4, but they soon became 14 – 17. They were in use before the additional four small cars which took numbers 10 – 13. As the Coal Company had been left to their own devices, they did not order these from the Brush Company, but went instead to Milnes, Voss & Co of Birkenhead. These cars had a rather unbalanced appearance with seven 'compartments' of 3 ft 11 ins width and one of 2 ft 7 ins – 18 inches shorter. It was rumoured that they were got at a bargain price, having been constructed for a customer who turned them down, but their cost (£3,500 for the four) would hardly suggest this. The original small cars, at £540 each complete, give an indication of the going price, and £875 for each bogie car hardly represents a great bargain. An old notebook has 'built for Cairo?' against them in pencil, but no evidence has been found to suggest that this is correct. Frustratingly, the contemporary ledgers which might throw light on this matter are not available. However, it is of relevance to note that the initial specification called for cars measuring 40 ft 6 ins long over the fenders; this dimension on the cars as built measured 40 ft 0 ins. It may be that they were in fact built longer and had to be shortened, but no conclusive evidence has been located. The first two were delivered to Leven Dock siding on 15th April 1907, but they did not enter service until about 10th June. Despite there being no particularly tight bends on the Wemyss line, perhaps initial trials proved they were over long (perhaps for safe passing on loops) and a judicious shortening had taken place ... It is unlikely that this can be satisfactorily resolved at this distance in time. They entered service with sides enclosed by removable panels, but there are no memories of the cars ever having operated as open. Appearance of the two parcels vans was also delayed; perhaps the need to re-gauge had not been appreciated either.

At this time, during November 1906, management of the tramway side of the undertaking was revised, with Knocker taking charge of the Coal Company's growing power supply department. A new manager, Major William T J Marshall VC, formerly with the 19th Hussars and latterly camp quartermaster at Aldershot was appointed. Probably unique amongst tram managers, the possession of the ultimate heroism award did not best qualify him to run a tramway undertaking, so ..." to learn his business a temporary manager will be employed, who will be a thoroughly qualified man ...". James Whyte (or White) from Birmingham was this appointee for a salary of £7 per week, taking up the post at the end of November and returning south before the end of March. Marshall earned his award for the rescue of his Commanding Officer during the Sudan campaign of 1884. "... his chief had his horse killed under him, and, severely wounded, was lying on the ground when the Dervishes closed in on him. Marshall forced his way to his side and kept the horde at bay till at an opportune moment he seized the Colonel and dragged him back to the British lines ...". He was also held in Ladysmith during the Boer's 110-day siege in 1899 – 1900.

For this new rolling stock a three-track enlargement to the depot was necessary, put in hand by Thomas Topping. One of Marshall's first decisions was to arrange for a maintenance workshop to be built in the depot, ending the necessity for Wemyss cars to be cared for by Kirkcaldy. Machine tools 'to the value of £220' were purchased. This occupied the end of lye 1 in the original part of the depot. The agreement to keep Wemyss cars overnight at Gallatown was also ended. The bogie cars were too long to negotiate the curves into the Kirkcaldy depot, and they did not participate in the through running arrangements. They were however, regularly, run down to Junction Road, where they could be turned carefully on the triangular junction, end for end, to equalise wear on bearings, wheel flanges etc. The small cars were also turned there when necessary.

In addition to the depot enlargement in hand by Topping, Aberhill Farm buildings were taken over by the tramway company in August 1907. Previously the company used a newly-built house opposite the depot. The animals were evicted from their byres; after a coat of whitewash these became the stores, with the two-storey house and other outbuildings refurbished as management offices. The manager's office was upstairs, overlooking activity below. All this construction was undertaken by the Estates' tradesmen, who also set to on the south side of the road to construct a clock tower, tea room, waiting room and shelter plus the 'Tower Tavern', all, including the new offices being in use by the end of the year. These

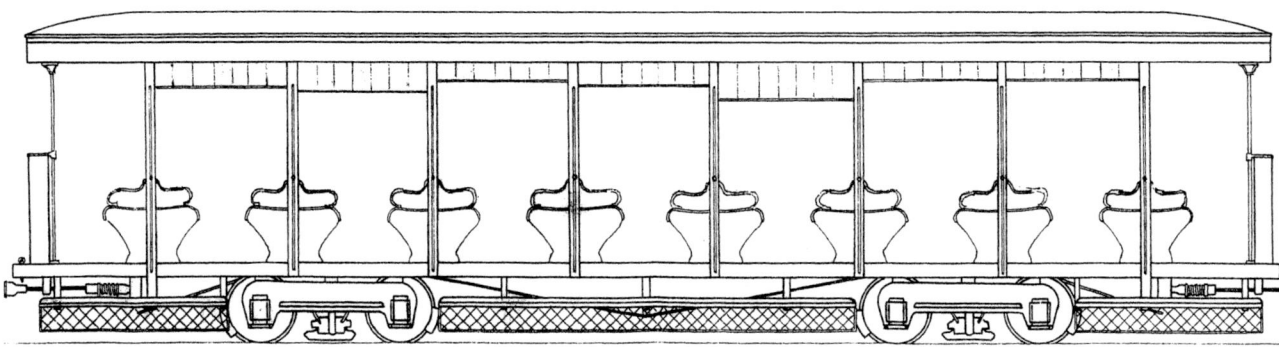

It had been the original intention to have the bogie cars haul trailers, to which, in July 1906, the Board of Trade very reluctantly agreed. Drawings, to the design shown, were prepared in Sellon's office, submitted and approved. In the event the proposal was abandoned without any record of just why.

At the start of February 1907 Major William T J Marshall VC was appointed as general manager of the Wemyss tramways. Over 20 years earlier the former Quarter Master Sergeant had rescued his commanding officer from being killed by Dervish hordes during the battle of El-Teb in the Sudan conflict, this act of extreme valour earning him the highest British military recognition.

Wemyss commenced running into Whytescauseway with no regard for the Corporation service already operating, this causing a considerable degree of friction. After taking advice from the manager of Leeds tramways, a workable system using cars of both operators began, under which the car mileage of each undertaking was balanced. This was the service advertised on 20th November 1907.

The Wemyss & District Tramways Co., Ltd.

THE TRAMWAY SERVICE HAS NOW BEEN COMPLETELY RE-ARRANGED.

THE first Car leaves LEVEN at 6.25 a.m., and from 8.25 a.m., on MONDAYS and FRIDAYS, there is a 20 Minutes' Service. From WHYTE'S CAUSEWAY, the 20 Minutes' Service holds good from 7 35 a.m.

On SATURDAYS, from 11.15, there is a 10 Minutes' Service from LEVEN, and from 12.15 from WHYTE'S CAUSEWAY

On SUNDAYS, there is a 20 Minutes' Service from 11.35 a.m., from LEVEN, and from 9.45 from WHYTE'S CAUSEWAY

W T J MARSHALL, Manager.

buildings were constructed at the expense of Mr Wemyss who then gifted them to the tramway company. The tower was modelled on the Tolbooth of West Wemyss; the clock was recognised to keep exceedingly good time and was used to regulate timetable operation. To complete the tower a gold painted swan weather-vane was added. The tram frequency operated by this date was generally every 20 minutes to Gallatown, with every second car proceeding on to Whytescauseway. From Leven to Gallatown 6d was the Company single fare, with the Corporation charging 1½d from there on to Whytescauseway.

The through running arrangements were, in practice, found to be far from ideal, with no coordination of the Wemyss service on the Kirkcaldy system. Also the Corporation were of the opinion that the staff of the Company cars were insufficiently aware of the fares and regulations while running on the town's lines. Eventually matters deteriorated to such a degree that Sellon was asked to appoint an arbiter to resolve the difficulties. He turned to J B Hamilton, manager of Leeds Corporation Tramways to find a practical solution. His proposals were implemented from 14th November, with an arrangement equalising the miles run by Kirkcaldy cars on Wemyss lines with mileage run by Wemyss cars on the Kirkcaldy lines. This was achieved, by the seven cars needed being five from the Company and two from the Corporation. The Saturday 10 minute frequency used ten and four cars respectively.

With his tramway now *fait accompli*, in December 1906 Mr Wemyss' restless mind moved to embrace another innovative project. He entered negotiations to acquire the British rights to a novel motor car incorporating mixed petrol and electric drive, the Mercedes-Mixte. Although he appears to have been advised that the company was an offshoot of the German motor works established in Stuttgart, that pioneer undertaking distanced itself from this newer enterprise, based in Vienna. In January 1907 a statement was issued advising that neither Mercedes Electric nor Mercedes-Mixte had any connection with the pioneer Stuttgart company. The principle in which he was interested was a novel one, but proved a dead-end at this stage in the development of the motor car. In these vehicles a 15 horse-power Mercedes petrol engine drove an electric generator with the current generated supplied to electric motors, one mounted in each rear-wheel hub. This arrangement dispensed with clutch, gear box, and gear lever. The simplicity of driving was a heavily emphasised selling point particularly with regard to ease of use by women. A

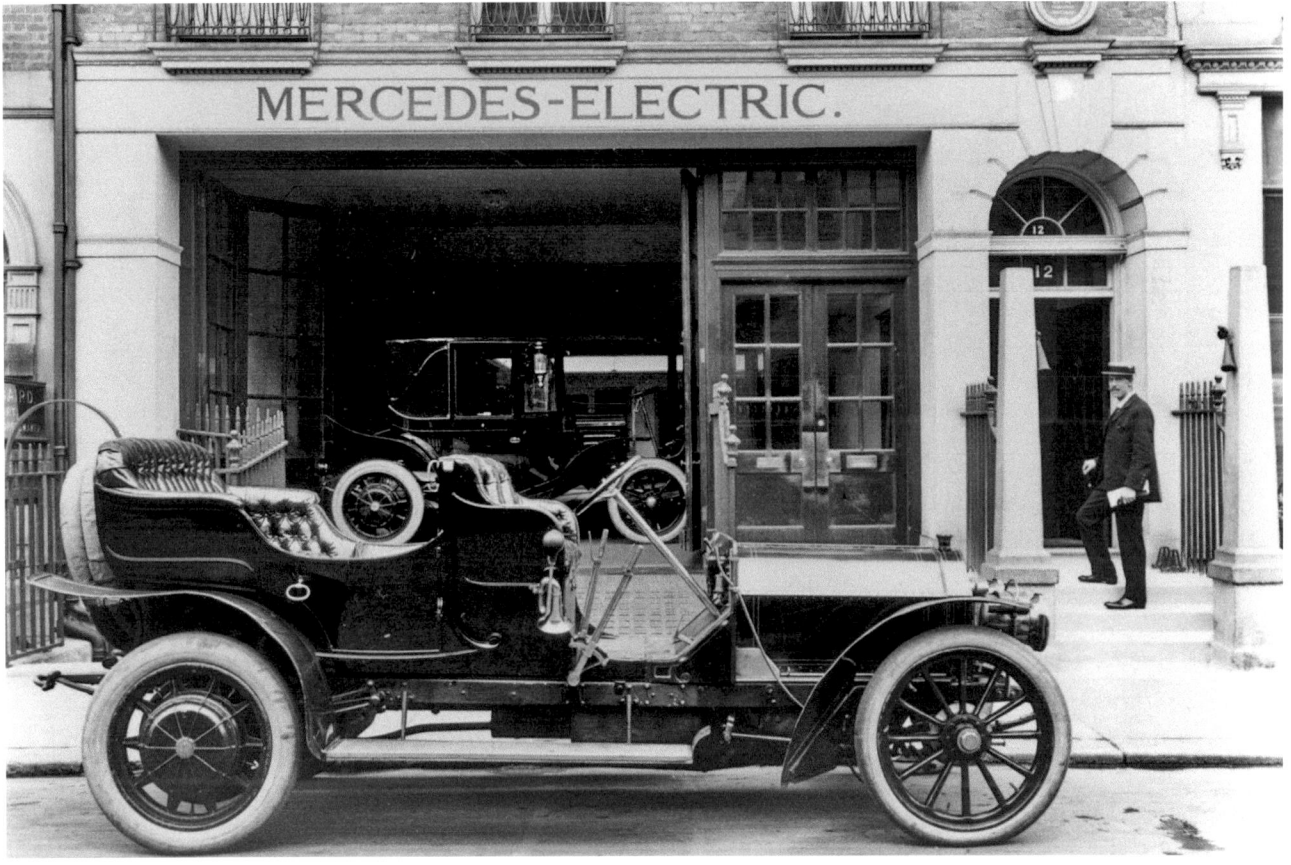

Mr Wemyss proposed a motor car construction plant for Methil and negotiated for rights to built electric drive Mercedes vehicles. An early prototypes is seen in London, but legal difficulties prevented the idea germinating.

site for construction of the factory was earmarked off Cowley Street, adjacent to the brickworks then under development. However, the association ended in April 1908 when Mr Wemyss acknowledged, no doubt with regret " ... they will, as far as the name is concerned, have sold us something which is not theirs to sell ...". Had the venture been a success, perhaps not only the economic future of the immediate area, but the entire developmental direction of the automobile could have been considerably influenced.

The new 'Wemyss Parcel Express' commenced on 1st May 1907. A parcels van was attached to cars on three daily journeys with parcel handling agents appointed throughout the immediate area. In the first six months of operation a profit of just £14.11s 3d was made. This did increase as the facility became more popular. The vans were found useful by the local constabulary on more than one occasion for conveyance of 'drunk and disorderly' to the cells at Buckhaven Police Station. The fare paid for this service is not known. However, it has to be recorded that the concerns of the Board of Trade appear to have been entirely justified. While on most tramways in Europe and further afield, the use of trailers was the norm, their use in Britain was the exception, and strenuously opposed by the Board. It was probably a result of this unfamiliarity which led to several incidents. The first, culminating in a fatal outcome, occurred just two months after the service started. An elderly man crossing the road after a car passed, failed to notice the van behind and was knocked over. At this time Kirkcaldy Council were in communication with the Board of Trade regarding the possibility of using trailers, but got very short shrift "... we have very decided views as to the danger of trailer cars ...".

Suffering from increasingly poor health, Mr Wemyss, following medical advice, in February 1908 went off to benefit from the invigorating climate of Menton on the French Riviera. But even during this time of indisposition he maintained an active interest in and control of the enterprises of his estates. He had been impressed during his overseas travels by the large size of rail wagons for bulk transport of commodities, particularly coal in his case, and was actively promoting introduction of 30 or 40 ton wagons for his collieries. This proposal found no favour with the rail companies or his fellow coal proprietors and the use of regular 10 ton wagons prevailed. The inordinate cost of alterations to all shipping hoists and similar installations was an obvious and major disadvantage.

From Menton he returned to his Kensington residence, where his health continued to deteriorate and he died there on 17th July, just days after his 50th birthday, mourned by many. It is interesting to speculate how he might have ensured the development of his Fife estates had he been spared for a longer life. He was determined to push construction of an extensive tramway facility throughout West Fife, for which Parliamentary approval had been granted. If outlying villages such as Kinglassie and Auchterderran been served by a tram network linking them to industrial and economic centres, the potential with a large mobile workforce on tap, was limitless. All of this, however, was nipped in the bud with his premature death; Fife had lost its greatest industrial entrepreneur. The directors recorded "... He was not only the pioneer, but also the promoter and Builder of the Tramway, and to his foresight, ability and financial assistance the boon of Tramway Communication has been given to this large and populous district. He foresaw the great possibilities in the way of public convenience and practical utility of such a scheme, and also the extent to which the travelling facilities thereby afforded would be taken advantage of ; and it was a matter of great pleasure to him that through his efforts and instrumentality the whole population of the district should possess the means of travelling with facility and comfort, and at the times suiting their own convenience ...". At this very time Mr Wemyss was actively pursuing the extension lines through the western and central mining areas of the county. Again his intentions were to take matters into his own hands; to float a separate company with capital of £100,000 then on his own account construct the first section, from Dunfermline to Bowhill. Then to extend this from Bowhill by Kinglassie and Thornton to join the Wemyss line at Gallatown. Negotiations with local authorities commenced but they were found excessively demanding. These bodies now had a practical demonstration in front of them of the benefits brought by the trams, and were anxious to achieve every possible advantage to themselves through road widenings and pavings etc. Mr Wemyss had had the great benefit of building through his own estates; now there were other, more rapacious, landowners and local authorities to satisfy. Fife Council wanted long lengths paved in granite setts, also that all lines on public roads had to be in the centre, rather than on the 'waste' to one side, as had been done for the Wemyss tram tracks.

With the benefit of a century of hindsight, it is probably just as well for potential shareholders that the proposals did not progress further. Despite the then financial success of the Wemyss line, it had in-built benefits which would not accrue to this new scheme, and even then had had difficulty in attracting investors. The balance in favour of tramways changed radically after the First World War, and the nature of the proposed routes meant that they could be just as easily be served by the motor bus, where rapid development had been achieved, and numerous surplus vehicles were available after the armistice.

By the end of the first decade of the 20th century the Wemyss tramway was an established and much appreciated part of the local infrastructure, regularly carrying almost two million passengers annually while providing regular returns for preference shareholders plus dividends (up to 7½%) for investors in ordinary shares. Wages had been increased with drivers' original hourly rates raised in 1908 from 4¾d per hour then to 5d, becoming 6d maximum (conductors' wage was 4¼d to 4½d per hour to 5½d). A decision had been implemented to replace track where mining subsidence had caused deterioration of the (unreinforced) concrete strip foundations used on track not originally laid on timber cross sleepers. It was also proposed to replace the roadside line along Randolph Road with a completely new reserved sleeper track in the field to the north. This included a new bridge for the tramway over the NBR's main line. Despite strenuous efforts by the Wemyss' engineers the leasee of the ground resisted all entreaties to part with the strip along his field. Thus when Randolph Road was eventually rebuilt as sleeper track (but still as single track) it remained in the original position on the 'waste' to the north side of the road. This work was undertaken by Wm Griffiths, the contractor who had laid the original track, on a cost plus basis (10% on labour, 5% on materials).

In October 1908 it was decided to relocate the much deteriorated track on Main Street Coaltown of Wemyss to a new sleeper track on the north edge of the village. The dangerous nature of the line here was also apparent, close to the footpath and passing in front of the village school. The accident potential came to pass shortly when a young lad, chasing a ball, was fatally injured. It has been speculated that this was the reason for the alteration, but, tragically, the decision had already been made. Additional works were also undertaken by Griffiths at this time, with several loops lengthened including at North Lodge, East Wemyss and Muiredge. Concrete strip foundations were replaced by sleeper track whenever renewals were being undertaken. Some £8,000 was spent on track improvements at this time, with the length of single line reduced in total by over a mile, a not inconsiderable proportion of the operated length of 7½ miles. Further track doubling after the First World War improved the situation until eventually 85% of the line was of high quality double track, all on sleepers, with fast running possible and journey times considerably reduced. When the Coaltown deviation was inspected by the Board of Trade, application was made for revision of maximum speeds. For considerable distances, mostly where reserved track was available,

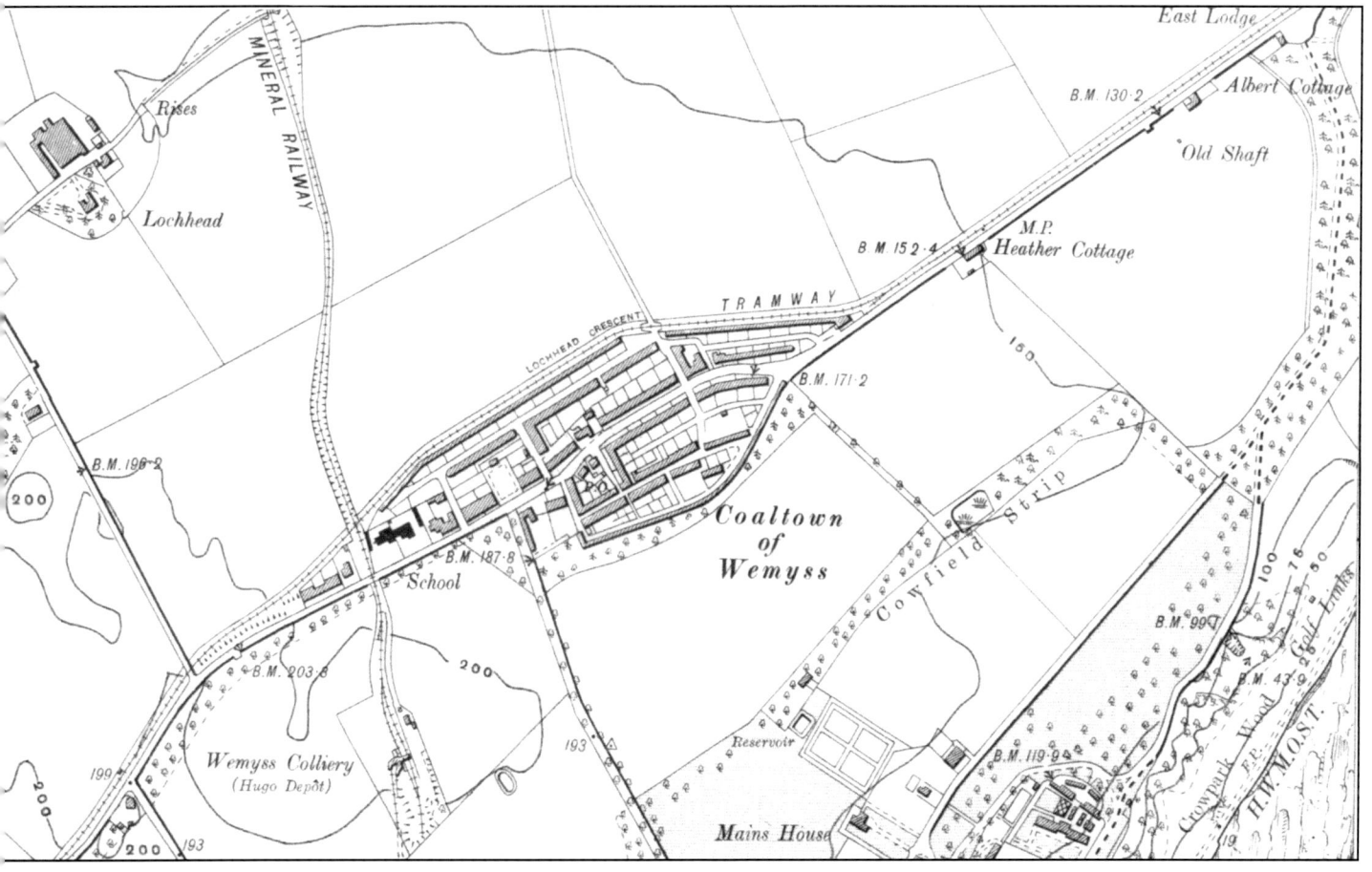

n of tramway at Coaltown, showing the diversion north of the village. Originally the line joined Main Street west of the mineral railway el crossing. It then ran through the village, to rejoin its original private track east of the 'Earl David' pub.

this was to become to 18 mph – regarded indeed as 'high speed' – it was not until 1930 that the maximum *permitted* speed for motor buses was increased from 12 to 30 mph.

An accident involving the parcels van occurred in January 1911. At the North Lodge a man attempted to jump on to a moving car, slipped on the hoar frost covered sleepers of the reserved track and had his legs badly damaged by the van. The fault was entirely his, but the Board of Trade's concerns were justified. On many tramways, particularly on the Continent, use of trailer cars was normal and passengers were conscious of their presence. The same did not apply here; similarly on modern British tram systems cyclists find difficulty in coping with tram rails in the street, while for example in modern Amsterdam cyclists and trams co-exist happily with no such problem.

In June 1911 Major Marshall intimated his retiral, with Mr Dawson then promoted to become general manager with effect from 7th October; he was to serve in this capacity for the entire remaining operational existence of the undertaking.

With demand for electricity increasing, and to allow the Denbeath power station to be utilised solely for local and colliery needs, negotiations commenced to transfer the Wemyss trams' power supply to the Fife Electric Power Company's (FEP) large power station at Townhill, just north of Dunfermline. Founded as an independent undertaking at the end of 1902, control of the FEP was acquired by George Balfour in 1906, immediately installed as managing director. He next formed the Fife Tramway, Light and Power Company (FTL&PCo) to hold the issued capital of the Dunfermline & District Tramways Company (DDTCo) and the Power Company. The Townhill generating plant supplied power to the Dunfermline trams whose first lines opened in November 1909.

Balfour commenced his engineering career in Dundee and with associate Andrew Beatty (an accountant) founded the eponymous partnership in 1909. The pair came together when they both worked on tramway construction in Britain for New York based J G White & Co – including *inter alia* the tramways of Dundee, Ayr and Southend-on-Sea. The new partnership seized the opportunity to develop power supply and associated tramway schemes with the contract for the Dunfermline lines being the first they tackled together, trading as Balfour, Beatty & Co Ltd. This was the foundation for probably the largest portfolio of power and tramway undertakings existing before the First World War. Not content with this, Balfour's political ambitions saw him as Conservative MP for Hampstead from 1918 to 1941 (having earlier twice, as a Unionist, unsuccessfully contesting the Liberal seat of Glasgow Govan).

Stephen Sellon considered that the new FEP supply should feed to a sub-station near to the Rosie Colliery at

George Balfour was a man of many parts. As founding partner of the Balfour Beatty Group of power and traction companies, he was on the boards of management of – at one time – no less than 33 such subsidiary undertakings. As if this was not enough, he also served as the Conservative MP for Hampstead from 1918-41.

Wellsgreen and believed that this had been agreed with George Balfour. He was therefore shocked to discover that the sub-station was already being built, but at East Wemyss, adjacent to the level crossing with the mineral line serving Michael Colliery. This shortened the length of the FEP's overhead transmission line connection considerably, but it still measured over 12 miles. At this time Balfour undoubtedly already had his acquisitive eyes on the Wemyss tram undertaking, and the need for additional power beyond the ability of the Coal Company's small generating plant came at exactly the right time. The new overhead feeder from the FEP's Lochgelly sub-station to East Wemyss consisted of six bare copper conductor wires (at 12,000 volts) and went live on 11th December 1911.

Soon followed a series of mishaps with much disruption of the tram service. By its overland nature, this new transmission link was highly susceptible to damage and the local miscreants enthusiastically took up a new game – throwing a handy length of fence wire over the wires, causing exciting short circuits! This behaviour was rapidly stamped out, but the new arrangements were prone to damage in high winds. The following winter was exceptionally windy, with several gales causing havoc bringing trees down on the exposed wires. Service was frequently interrupted, on occasion for several days, causing major disruption. Some power was obtained from Kirkcaldy Corporation, by joining the two systems electrically at Gallatown, with a mile length of the Wemyss' overhead eastwards replaced with heavier section wire.

Meantime Balfour was busily buying up all the Wemyss shares he could get; when he became the majority shareholder he was given a seat on the board. Wasting no time, he took the role of chairman from the first board meeting of 1912; at the same meeting Messrs Sellon and Budge resigned. Of the other directors, Mr Oswald and Major Gordon immediately retired followed by Mr Bowman in May, leaving W Ogilvie Shepherd as the only survivor from the previous regime. Mr Dawson remained as manager, with enhanced salary, holding this position for the entire remaining duration of tram operation by the company. Advertising on the cars had, from the start, been let out to a Mr Abrahams, but he was continually in arrears with payments. Consequently his agreement was terminated in March 1912 with an instruction given for his boards to be removed from the cars. A new contract was agreed with the Sanjebord Publicity Company with just one board on each side of the cars' roof – Abrahams had boards all round the roof edges. This rather altered the appearance of the cars, until at the end of 1920 when this contract also ended and all external advertising boards were removed.

Balfour continued investment in replacing and improving the tracks, with a programme of doubling wherever possible. With the Board of Trade allowing an 18 mph top speed on the reserved track, an improvement in running time was possible, with an overall reduction of fifteen minutes. Passenger numbers, in these pre-First War years, increased regularly to reach 2,135,682 in 1913. The figure produced by Stephen Sellon for anticipated passenger numbers, and used in the prospectus, was for a population of 70,000 making an average of 30 journeys annually; it took seven years to achieve this. Despite the considerable cost of improvements, investors in ordinary shares – including those locals who had not sold off their holdings to George Balfour – benefitted from regular dividends; usually 6%, but reaching 7½% in 1913. Balfour Beatty extracted an annual fee of £500 for their "expert management". To the travelling public the most obvious change saw new uniforms for platform staff and from April 1913 a change in the colour of the cars. Away went the Wemyss 'livery' estate colour of mustard yellow (almost mimicking American 'traction orange' much used there for street cars and buses) to be replaced by a rather more sober hue, officially described as 'Mahogany Lake'. However the affectionate nickname of 'mustard boxes' remained with the cars for the rest of their days. Samples of these paint colours retrieved from car 16 in the 1960s reveal that the closest matches in standard colour identification today (BS381C) would be 363 for the yellow and 449 for the lake. However both colours were affected by grime and weathering and by how often (or seldom) they were varnished.

When the original tramcars were purchased each was fitted with two motors made by Bruce Peebles, each of 25 horse power. These were hailed as being of the latest and best design, with outer casings in two pieces which could be split horizontally for ease of access for routine maintenance. However, it was found in operation, that this joint face wore and became difficult to seal to keep oil in and water out. Mr Dawson recorded "... the old equipment is a constant source of worry ...". Five sets of replacement motors were supplied in 1914 by Bruce Peebles, but the other four of the earliest cars were then fitted with motors purchased from West Ham Corporation Tramways. Four cars soldiered on with the original equipment until in 1922 these were replaced by second-hand sets from Liverpool Corporation trams. A shelter for passengers was built at College Street, Buckhaven in 1914, adding to those already provided at Aberhill, East Wemyss and West Wemyss road end. Another was also built at Gallatown terminus at the end of 1914, the cost split between the Company and the Corporation.

The consequences of events in Sarajevo on 28th June 1914 set in motion a tragedy which eventually led to the First World War with dreadful and totally unanticipated effects reaching almost every family in the land. The initial display of nationalistic fervour led to many of the country's finest and fittest men volunteering for patriotic service and the Wemyss Company, like many others, felt the effect of this depletion almost immediately. The parcels express service ended at the end of November 1914, never to resume. The little 'monkey vans' as they were known locally lay beside the depot for years with one later to be found behind one of the Wellesley Road houses near the school.

In September 1916 Mr Dawson employed the first female conductors. They became a permanent feature, unlike on some undertakings where this role reverted to male employees after the end of the War. In the late twenties, after the expansion of the company into bus operation, there were approximately equal numbers of male and female conductors, the latter working mainly on the buses. Some women were trained to drive trams and a direct benefit of the change was said to be that male behaviour – particularly on workers' cars – was noticeably improved. It was also said that the 'lassies' on Wemyss buses had a reputation for good looks and demeanour; being in the public eye led to more than one romance.

At the beginning of the war, use of the trams declined, this trend soon reversed as more and more workers travelled to work, drafted in for employment associated with the 'war effort'. Numbers of passengers carried increased, reaching 3,691,552 in 1920. Dividends reflected this unanticipated bonus, reaching 10% at this period. Difficulties were experienced with the through running service from Leven through Kirkcaldy into Whytescauseway. The Corporation single line between Gallatown and Junction Road was a frequent cause of delay, and while the obvious – and not unduly costly – remedy would be to double the track, the Corporation was unwilling to invest in this for what they saw as a benefit to the Company only. The Company also was restricted by being unable to use the bogie cars on Kirkcaldy's lines as their curves were too sharp. To cope with passenger demand at busy times it had became necessary to operate the small Wemyss cars duplicated; costly in manpower and power. To allow greater flexibility, through operation on Saturdays ceased in November 1917, and with the benefits gained (if not for the passenger) all through running ceased after 15th January following. In each week of 168 hours, a tram service was operated on no less than 140 hours causing strain not just on staff, but also on vehicles. Part-time drivers were recruited; several miners would (it was said) do a day time driving shift followed by a night shift down the pit. Not to be recommended. Despite the end of through running, the Wemyss cars – including the bogie cars – were still regularly taken into Kirkcaldy to Junction Road to be turned end for end on the triangular junction to equalise wear, there being no such track on the Wemyss line to permit this.

That the company intended to see the trams as the prime form of transport between Kirkcaldy and Leven was confirmed when, soon after cessation of hostilities, work resumed on track doubling where left off. North Lodge to Claydales was first tackled, followed by Coaltown to East Wemyss, then East Wemyss to Rosie and finally Rosie to Muiredge. This represented the last such work undertaken, with more than 80% doubled. It had been hoped that the line along Randolph Road would also be done, but this could not be agreed with Fife County Council and the Earl of Rosslyn's Trustees, while at the other end of the line, the narrow nature of Leven High Street prevented any improvement there. The need for additional rolling stock was also apparent and in January 1920 Mr Dawson was instructed to open discussions with Balfour Beatty to obtain four new cars – the largest modern cars which could be obtained on 8 foot wheelbase trucks. It may be that this was with a view to resumption of the service into Kirkcaldy, but it is probable that long wheelbase trucks would have presented the same problems as the bogie cars.

These proposals suffered a severe setback when a strike by miners during three months of 1921 brought the local economy, dependent to a very large degree on this trade, to the edge of ruination. As a wartime measure coal mines had been taken under government control in 1916, then returned to their owners in March 1921. A post-war slump in demand led to a proposal to reduce wages, resisted with strike action mooted by the 'Triple Alliance (rail workers, dockers and miners). This commenced with mines ceasing production on 1st April, but was not then supported by other parties. Mine workers held out until 28th June when, without

The first 'lady car conductors' were employed in August 1915 as a result of the large number of tramway staff volunteering for war service. This uniform has an embroidered badge, and the initials 'W' on the right lapel and 'T' on the other.

achieving their objectives, they resumed their labour. Coal supplies for power stations were much depleted, generating capacity affected, and tramway operation perforce reduced. The seeds had been planted for the General Strike five years later, but also the dependability of the tram service had been found susceptible to the availability of power from just one power station. This was a typical situation prior to establishment of the National Grid in 1935.

The national electric power generation situation as realised during the First World War (with over 600 generating plants producing power at various assorted voltages), was obviously totally unsatisfactory, and became the subject of the 1919 Williamson Report to government. This in turn led to a commission under the chairmanship of Lord Weir which in 1925 proposed formation of a national linked network (gridiron) of power transmission at standardised voltages. The Central Electricity Board was formed to build and operate this National Grid and control – but not own – power stations. Mostly above ground on architect designed pylons, this scheme aroused great controversy, but created a large number of construction opportunities at a time of economic depression and unemployment. The number of generating facilities was reduced to 132 – but still including that at Townhill. This was owned by the Fife Electric Power Company, a division of the Fife Tramway, Light and Power Company, both subsidiaries of Balfour Beatty Limited. Townhill still supplied power to the Dunfermline Tramways, yet another part of the Balfour Beatty empire.

At this time it was still the intention to construct tramway extensions for which powers had been obtained. To rationalise this, the powers to build tramways numbers 2, 2A and 3 which had been obtained by the Wemyss Company in 1910, running to Auchterderran, Cardenden and Kinglassie from Lochgelly were transferred to the Dunfermline Tramways. For this consideration the Wemyss undertaking received £1000 worth of shares in the FTL&PCo. The economics for the final link, joining Kinglassie to Thornton and on to Kirkcaldy and Wemyss were not sufficiently enticing and these powers were allowed to lapse.

The effects of the 1921 coal strike, allied to the number of unemployed, plus the large numbers of motor vehicles being sold off as war surplus, gave rise to a climate where budding transport entrepreneurs could, and did, with a loan to purchase a bus or charabanc then try their luck, either by running a regular service or by offering tours to one of Scotland's many beauty spots. The first such 'pirate' operator to go 'head to head' with the trams were the brothers Anderson, Andrew and Tom, from Buckhaven. They traded as the 'Buckhaven Motor Carrying Company, their first bus being a small 14-seater named *The Pioneer* based on the ubiquitous model 'T' Ford . With this they mounted a challenge to the established regime by running from Leven to Gallatown 'chasing' the trams, commencing in May 1922. This competition was not achieved harmoniously and numerous appearances in the local Sheriff Court rapidly followed, with buses obstructing trams, and vice versa. Tempers flared and verbal and physical violence was, unfortunately, resorted to repeatedly. Court appearances by those involved were also frequent, with the guilty party usually fined £1 – with the option of 7 days imprisonment.

Anderson was not the first to operate buses in the area. In April 1920, John Dewar (who had just taken over the garage which had developed from the posting establishment attached to the Caledonian Hotel in Leven) gave up the horsing element with the animals and plant auctioned off. Before the end of that year he had purchased his first charabanc and was running a successful regular service to Anstruther. With additional vehicles he joined the expanding gang of chasers to Gallatown, following the tramway, but diverting via Buckhaven. While moderately successful, he offered little if any time benefit; while the ride in a solid tyred vehicle was distinctly inferior to that provided by the trams. His bus business was formalised as the Caledonian Motor Carrying Company. These earliest bus operators often started with just one vehicle, tried a route, and if not profitable, moved elsewhere. Hence their operations were not always recorded, particularly if they did not advertise their intentions; it is thus difficult sometimes to be specific about who was first to operate where. Some memories of those pioneer operators were recorded in the 1960s – unfortunately often adding more confusion to the record.

These 'chasers' were a source of revenue loss to the tram company, but management rapidly decided to fight fire with fire, and in June 1922 purchased three 'light buses'. These Tilling-Stevens petrol electric were of a design which other Balfour Beatty group companies had adopted as their preferred choice. Significant numbers of this design were purchased by the omnibus department of the Falkirk tramways from 1914 on, then Wemyss and echoed by Dunfermline tramways from 1924. Unfortunately as the chaser operators developed, they opted to buy faster, light weight vehicles which could show a clean pair of heels to the more ponderous Tilling-Stevens. Bodywork on the BB group vehicles tended to be from Strachan & Brown of North Acton, London. Eventually the bus department of the Falkirk tramways (which rapidly outgrew its parent and was then renamed the 'Scottish General Omnibus Company') owned nearly eighty of these; the Dunfermline tram-bus department had nearly twenty but Wemyss a mere dozen. To create an effective response to the chasers Wemyss then changed policy with purchase of Halleys then Albions.

The tram side of the business was far from being neglected at this time and track doublings recommenced which had been held in abeyance during the late War. Ownership of the fields on the north side of Randolph Road was now with the Board of Agriculture, and since the resistance of the former leasee was no longer an obstacle, the Wemyss company purchased a strip of land to allow for construction of double track on reservation. Negotiations began with the railway authorities for a dedicated tram bridge over the main line to Dundee. This was still being debated at the end of 1925, but other events ensured this new track was never built.

Mr Dawson was instructed to purchase four 36 seat, 8-foot wheelbase, four-wheeled cars. An order was given to English Electric in September 1920, but for just two vehicles. What happened to this decision is not on record, but the cars never materialised. Instead in January 1924 an order was placed with Brush for two long 45-seat bogie cars of modern design. These cost £2,000 each; this at a time when a bus would cost around half this amount (typically, £715 for chassis and engine and £315 for a Strachan & Brown body). The working life of a tram was expected to be more than twice that of a bus. These two cars were delivered during the first week of January following. Also at this time property was purchased at Gallatown where a new bus garage was proposed with accommodation for trams. This did not materialise.

Bus competition generally was becoming ever more problematic, and to deal with one of the original operators, at the start of April 1925, control of the Caley bus company was achieved by an agreed share purchase. Subsequently the efforts of the Caley were directed more – but not completely – to private hires, taxis and school runs. This was the first move in what became a resolute policy by Balfour Beatty to maintain their transport superiority locally. Just six months later came a move, instigated by the owners, to enquire if there was any interest in the Group making an offer for the General Motor Carrying Company of Kirkcaldy (GMC). Negotiations proceeded rapidly and harmoniously, with control transferred at the end of November 1925. With this came 83 assorted vehicles, including lorries owned by Scottish Utility Motors (a subsidiary of GMC). These companies retained their trading identities, as did others absorbed subsequently. In the case of the GMC the two founding partners (Alex Sturrock and John McGregor) were retained to manage 'their' company, generating considerable goodwill.

Probably the first Leven-based omnibus owner was the Caley Motor Engineering Company who advertised their charabanc as available for hire in April 1920. This might be that vehicle; while the owner's name is painted on the chassis. Appropriately, it is a Glasgow-built Caledon vehicle.

The travelling public saw little evidence of change, but items of major expenditure were now ratified by the Wemyss directors. GMC paid Wemyss £1,000 per annum as a management fee. Additional capital was raised at this stage by the Wemyss Company to finance these acquisitions, and for additional vehicles.

Trade throughout the country was completely disrupted by the first ever General Strike – of all workers – called by the TUC to support miners, which started on 3rd May 1926. Many workers rapidly drifted back to work and the strike was officially terminated after just nine days, on 12th May. Again the miners held out, until, as seen in a historical context, they were starved back to work at the end of November having gained little, if anything. Local transport services ceased completely for the nine days of the strike and revenues in all departments of the company were much reduced. At the Company Meeting in March 1927 it was recorded "... while the coal stoppage alone would, in itself, have been sufficient to produce unfavourable trading results, we have had at the same time to face exceptional operating difficulties arising out of intensified omnibus competition ... the effect of the unprecedented operating conditions is apparent from the traffic receipts on the tramway side which decreased to £14427 against £22198 the previous year ...".

This state of affairs put the future of the tramway, previously recognised as the 'backbone' of the company's operations, into severe doubt. Strenuous efforts continued to be made to reverse the situation with regard to the unregulated unprincipled and unbridled bus pirating. The situation was recognised by the various local authorities, which, led by Fife County Council, introduced a compulsory bus licensing and regulatory requirement at the end of September 1926. Probably already too little, too late. While the intent was entirely laudable and understandable, a considerable number of buses became superfluous to needs, as were the crews operating them. Bus timetables were rationalised and times allocated to the different operators. These applications applied on heavily contested corridors, the most trafficked of these being the tram routes with their 'chasers', Dunfermline to Lochgelly, and Kirkcaldy to Leven. On the first mentioned there were no fewer than ten bus owners (eleven if the tramway is included); just three on the Leven run. The timetable produced for Lochgelly was a masterpiece of ingenuity, but completely inoperable, leading fairly

One of the Company's Tilling-Stevens buses in the picturesque setting of Drummochy Lower Largo. The narrow bridge at the Harbourhead spans the Keil Burn. Known as the birthplace of Alexander Selkirk, inspiration for Daniel Defoe's classic tale of *Robinson Crusoe*, the area has lost none of its charm through the passing years.

quickly to wide scale abuse and/or evasion. The legality of the bye-laws was tested in the High Court and proved legitimate. The intent was admirable, but execution was impractical and resulted in little improvement, only in an increase in prosecutions for infringement, frequently relating to buses running ahead of timetable.

Absorption of competitors continued, next to succumb being A&R Motors, then Fuller of Newburgh, permitting a Wemyss service to be run from Leven to Perth via Auchtermuchty. For the first time, the adverse trading conditions resulted in no funds being available to pay any dividend.

At the end of 1928 the three Scottish tram companies in the Balfour Beatty Group attempted to achieve Parliamentary legislation to give them a high degree of protection including power to "... restrict or prohibit the running of omnibuses by any local authority, company, body or person along the routes of the existing tramways of the Company ... and restrict the licensing of any such omnibuses as ply for hire along such routes." Perhaps unsurprisingly, severe opposition from bus operators resulted, and while the first Order dealt with (for the Dunfermline Company), was passed, it was in a much watered-down form, with the critical protection clauses deleted. That for Falkirk was dealt with in the same manner whereupon the third, for Wemyss, was withdrawn. Kirkcaldy Council had refused to allow any bus operation within the Burgh until one operator (Smith Brothers) tested this assertion. On 21st December 1927 they commenced a 'Town' service in direct violation of the Council's edict. It soon became apparent there were no powers to prevent this, and the entire situation relating to bus use in the town changed. All out of town bus services hither to transferred passengers to the town's trams at Gallatown or Links termini, but now these services ran through to Sands Road (the Esplanade) picking up passengers in the town en route. The monopoly asserted by the municipal tramway had proved to be nonexistent. Mr Francis, manager of the Kirkcaldy system was required by his employers to produce a report on the future of the town's lines and he in turn wrote to Mr Dawson to ask what the plans were for the future of the line to Leven. This was a question to which Mr Dawson would dearly like to have known the answer.

Despite these unsettling events, in early 1929 it was decided to purchase two second-hand trams to augment the fleet. A decision was made – probably by George Balfour rather than Mr Dawson – to purchase two tram bodies from the Potteries Tramway system based in Stoke-on Trent. Cash books now available give detail of the transaction. Purchases were through E J Walsh & Co of Edinburgh who dealt extensively at this time in 'recycling' spares and tramway fittings from recently closed lines. Walsh had been manager of the Musselburgh tramway company for three years from June 1916 until December 1919, thereafter setting up as an electrical contractor in St. James' Square, Edinburgh – close to the then head office of the Corporation's newly electrified lines. In May 1929 Walsh was paid £120 for two tramcar bodies. And that is exactly what arrived – the bodies had been stripped of all electrical equipment, and came without bogies or any other mechanical gear. The original Potteries bogies would have been of little use as the gauge there was 4 ft 0 ins. The cars arrived in May 1929 and a steam crane was hired (£11.14s from the LNE Railway) to unload them. It has been thought that these were former Potteries cars 82 and 95 (dating from 1900, but subsequently upgraded) but there is now considerable doubt about this. Electrical equipment was purchased, again through Walsh; BT-H type B18 controllers, GE58-4T (37 hp) motors, trolley bases, trolley poles and trolley heads – all second-hand and possibly from the Potteries Company also. The cars, which were without drivers' vestibules, had these fitted at Aberhill using timber from Brown's sawmills at Buckhaven; even the glass used in these was second-hand, reclaimed.

Four new equal wheel bogies were made for these cars, with new castings made by the National Steel Foundry (NSF) of Kirkland, Methilhill. The design of these was as the 1907 Mountain & Gibson bogies supplied under workers' cars 14 – 17. Originally fitted with track brakes, these had been removed over the years. A bogie from one of these cars provided the original castings which were used to create patterns from which new items were cast. As modified at Wemyss these bogies had proved more reliable in service than the Brush design maximum traction type supplied with cars 18 and 19 in 1925. So much so that when these latter two cars went to the Dunfermline tramways in January 1932, these newest, locally manufactured, bogies went with them in preference to their Brush originals. Wheels and axles for the new bogies came from Hurst Nelson of Motherwell while Timken roller bearings were incorporated. In this rejuvenated state these already thirty-year old cars gave good service until closure of the line. The cars were repainted in Wemyss livery of gold-lined 'Mahogany Lake' which does not tally with the memory of one former employee who remembered the two 'English Cars', as they were known, as being of a lighter hue than the others. Coachpainting of cars and buses was always undertaken by J C Rolland jnr. of Leven; the tramway company had no paintshop facility as such.

The effect of bus competition on the country's railways was serious and dealt with by them quietly purchasing into the industry, in Scotland through the Scottish Motor Traction (SMT) group of Edinburgh. These through various holding and operating companies proceeded to mop up individual bus proprietors. Locally, this being effected via W Alexander & Sons (WA) of Falkirk. In February 1929 SMT acquired, by purchase, two-thirds of the share capital of Alexander, and before the year was out, a truce was declared between SMT and its subsidiaries, and the bus operating wing of the Balfour Beatty group. Early in 1930 this was extended to be an outright takeover, when for £450,000 (about £30 million at current values) all Balfour group bus operators were sold.

Possibly as a consequence of the failure to achieve protection for their tramways, the Balfour group, in mid-July 1929, decided to separate the power generating side of the business from the transport companies. The Fife Tramway Light & Power Company was sold to the associated Scottish Power Company for £1,325,000 (£87.5 million 2023) Their Scottish tramways at Falkirk, Dunfermline and Wemyss were still receiving investment, but did not sit comfortably in a portfolio of bus operations. Unlike most similar undertakings throughout the country, which were becoming run-down through lack of investment in up-grading and track modernisation, these lines all had considerable infrastructure improvement. At Falkirk the entire track had been reconstructed and a fleet of new trams put in service in 1929-30; much of the Dunfermline track had been rebuilt and at Wemyss track doubling continued into the 1920s and new trams had been added.

On the Wemyss line, the service was now run mostly by high-capacity bogie cars; the original 'mustard boxes' were still in service, but were probably approaching the end of their useful lives. The new regime had no interest in maintaining what they saw as competition creaming off passengers on potentially the most profitable routes. Investment in new buses continued, and the probably predictable conclusion seemed inevitable after an offer from Alexander for the Kirkcaldy trams was accepted by the Town Council. These were replaced by buses on 15th May 1931 amid riotous scenes. At this time control of the tramways element of the Wemyss undertaking was finalised when Alexander's offer to purchase all issued shares for a total amount of £19,752,3s.5d (£1.3 million) was accepted. It was the intention to close the Wemyss tramways immediately, but negotiations with local authorities were protracted and the matter was not resolved easily.

For the time being therefore, the Wemyss cars continued running and, in two unexplained episodes, once in September 1930, and again in May 1931, Walsh was despatched at the Wemyss Company's expense to Birmingham, to examine used tramway material for sale at Tividale works in the West Midlands. Near Dudley, in the West Midlands' Black Country, this was the heart of the extensive 3 ft 6 ins gauge tram network operating west of Birmingham. Here the Birmingham & Midlands Tramways Joint Committee had run a network of several tramway companies, the last of which closed on 1st March 1930. Their works had built many cars for their undertakings, including over forty single deck vehicles, and it appears that Walsh was sent south to see if there was merit in purchasing any of their redundant trams, the newest just ten years old. The benefit of 'ready to run' second hand purchases was unfortunately outweighed by the transport cost, and although no complete vehicles were purchased, some spare parts were bought. (It is relevant to note that while the two Potteries bodies cost £60 each, transport by rail was just over £60, for two.)

A need for more trams was still considered, either to justify the number of bus licenses which might be needed in lieu of trams, or because the oldest trams were now past their sell-by date – or perhaps both. In May 1931 a more obvious source, closer to hand was apparent, and although Kirkcaldy retained eleven trams after closure of their lines for their own use as shelters etc, they still had fifteen trams in full working condition available for disposal. Four of these, probably the newest cars (numbers 23-26) built in 1915-16, were transferred to Aberhill in May 1931 where they were adapted for the Wemyss lines. Stairs and top-deck seats were removed and renumbered 22-25, and with the new owners name on the waist panel, they were in service by July. It seems probable that number 26 became 22, the others retaining their original identities. Four more cars, original numbers unknown, followed, put in service in October as Wemyss 26-29. These were older Kirkcaldy cars and needed more work done before entering service; stairs and top-deck seats were removed, also these cars required drivers' vestibules to be fitted to the platforms. Timber came again from Brown of Buckhaven and second-hand glass was used. Trolley retrievers were fitted, probably amongst the material got from Tividale. Other spares obtained from Walsh at this time included trolley wire, trolley heads, brake blocks and gear cases, probably also from Tividale. During 1929 Walsh had supplied a rail grinder (for £25), but no details are provided.

At the start of 1929, when the staff was at full complement, there were 172 employees on the payroll. These were: traffic superintendents, 6; inspectors, 5; motormen (tram drivers) 28; conductors, 27; cleaners, 6; track workers, 7; electrician, 2; car workshop, 10; bus drivers, 28; conductresses, 25; bus repairs, 18; cleaners, 9; greaser, 1. (Not including office staff.) With activity reduced in the final week (tram operation only) staff was reduced to traffic superintendents, 3; inspectors, 4; motormen, 15; conductors, 15; track workers, 5; electrician, 2; car workshop, 3; cleaners, 3; total 50 – again excluding office staff.

To avoid the riotous scenes which accompanied closure of the Kirkcaldy trams, little advance notice of closure of the Wemyss line was given, neither to public nor to staff. It was reliably anticipated that the last Wemyss tram was to run on the final day of 1931, but a reduced service was run during January, using bogie cars as the small cars were already being sold off as conveniently sized hen houses etc.. Disposal of these former assets was in the hands – not of Walsh – but of Dan Murphy, the former permanent way foreman. All did not go entirely smoothly, with a disposal of three cars as hen-houses ending in court. Three related farmers at Fedinch, Monksholm and Caldside each wanted a car, at £15 (£42 for three) plus £5 each for transport. All went well, except that the first mentioned proved impossible to deliver thus was not paid for; it ended at Lathones. Murphy sued for his money but it was judged that as no contract existed the case was assoilzied [discharged]. For many years the old vehicles could be seen throughout East Fife. Of the workers' cars, one became a men's club house on the Ness Braes at Buckhaven, number 16 was an unusual 'attraction' at Falkland village while number 14 was for many years a hen house at Blacketyside Farm between Leven and Lundin Links. There was an intention that this would be the subject of a preservation scheme, but this comprehensively failed when the owner put an end to matters by cremating it without further debate.

Eventually Saturday 30th January 1932 was intimated as the last day, with crowds gathering all along the route, but there was still no advice of the time of the final run. As the cold January evening wore on most people lost interest and wandered homewards – until, without warning or ceremony, the 10.15 pm from Gallatown ran its journey to Aberhill only, reversed into the depot, becoming the unheralded final car. It had been anticipated that Alexander's would offer employment to tram employees, but in the event only seven individuals were retained. Mr Dawson took a post with the Fife Electric Power Company before soon retiring to live in Largo.

On that last day of operation no less than 4,282 passengers were carried, including 894 using workers' fares; but only 14 tickets were issued for the end-to-end run. The Company remained in existence – possibly to satisfy a legal obligation of the contract to transport miners at shift times – until wound up in July 1938. Comprehensive details of every known vehicle, tram and bus, used during the thirty-plus years of its existence are given in the PSV Circle's publication SFE1.

Car 16's second life as a greenhouse in a Falkland village garden was nearing its end in this 1960s scene. However, the owner did allow brass fitments to be removed from the old tram before it was finally broken up. These have since been incorporated in other tramcar restoration projects.

An Aberhill Argument

The Tower clock had just struck twelve,
And twelve was late enough;
This midnight hour the fates decreed,
The weather would be rough.

Seeking shelter from the storm,
Close by the tramway shed;
I envied all the Methil folk
Now safely snug in bed.

Although I thought myself alone,
I heard a murmuring din,
Of voices somewhat strange to me,
Coming from within.

Strange to hear an argument
From a depot in the dark,
Distinctly now I overheard
A bus make this remark.

"Poor tramway cars submit to fate,
When people want a bus;
You've got the sack, it could be worse,
So why make all this fuss.

Think you of the bygone years,
Unrivalled was your sway,
From Leven town to Gallatown,
And most times made it pay.

Yes, my old decrepit shaker,
It's time for you to go;
Modern age must travel fast,
And you are much too slow.

However, when you finish up
Be sure I wish you well,
And when you're packing up your kit
Take the rails as well.

Don't forget the bolts and nuts,
And take the sleepers too;
Leave the road, I'll be content
To say farewell to you."

Then a sparking voice replied,
It sounded like a "damn,"
And other words unfit to print
In the language of a tram.

In angry tones it volleyed forth:
"You polished ruddy pet,
Dolled up in rich upholstery,
Your taunts I can't forget

The perfume that you leave behind
When you have passed me by
Is worse than Leven river
In the middle of July.

Go, you rubber highwayman,
Gloat about your speed,
Your mission for the people
Will develop into greed.

I might be slow but always safe
I fill no heart with dread,
When my journey is completed
I leave no trail of dead.

My service has been faithful,
And will be to the end,
To finish up as I began -
The miners' greatest friend."

Wm J Blair Leven 1931

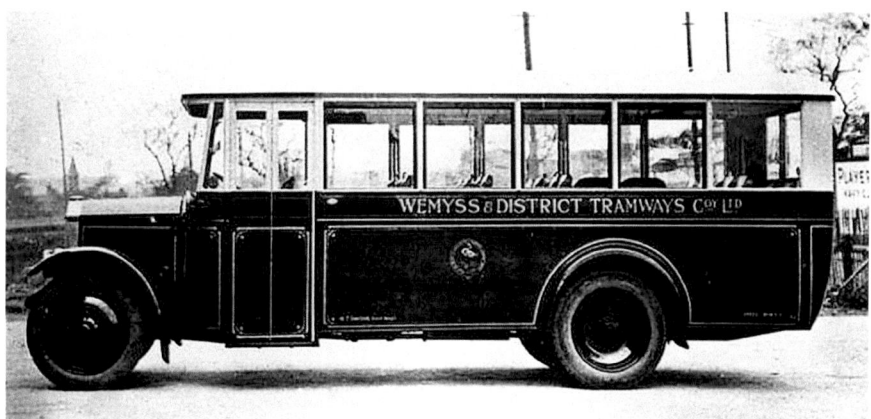

Albion PJ26 25-seat bus of 1928 (probably FG 4117 or 8) photographed at the Acton works of body builder Strachan & Brown (Their builders plate is to the side of the front-entrance door). Strachan & Brown supplied many of the Wemyss' bus bodies; these on average cost £275; the chassis with engine, £750. In 1928 the Wemyss Company licensed 28 buses.

Car 8 leaving the west end of Coaltown Main Street heading for Gallatown, with a full load on board and under the gaze of the local children. The proximity of the line to the ill defined footpath is only too obvious.

At the eastern end of Main Street, car 6 is just as much the centre of attraction. The car is filled with eager joy-riders and local children are fascinated by this new phenomenon. The line was diverted away from Main Street, the dangers it presented being apparent; the deviation opened in September 1909.

Cars 2 and 4 pass at the first loop on the private track east of Coaltown. Behind car 2 is the site to be occupied by the 'Earl David' public house. Opened in 1911 it was conceived as a charitable project by Lady Eva Wemyss, and operated on the 'Gothenburg' principle. It closed as a public house in 2017.

Another gathering of excited children to admire the new trams, here at the next village along the line, East Wemyss. In the foreground is the mineral railway level crossing with the line which served the very productive major employer, the Michael Colliery. Behind is the 'Car Shed' waiting room.

Although two weeks' operation had passed, the trams still exerted a fascination – and not just for the children. Car 8 has paused at the 'Car Shed' stop, perhaps waiting for a car travelling to Leven to arrive from the single track, then allow number 8 to proceed to Gallatown.

The 'Car Shed' featured in bus timetables until recently, and for most of its life served as a popular newsagents. The rustic tree-trunk pillars supporting the canopy are a design 'trademark' of the estate architect Alexander Stewart Tod, and can be seen on local estate cottages to this day.

Car 8 showing how the track was built on the road 'waste', effectively creating a private right of way there. Here it is just about to join the fenced off reserved track. To the left, the red sandstone ruin of Macduff's Castle, probably dating from the 14th century. It was still inhabited three and a half centuries later, but then was allowed to become ruinous.

A few yards to the east the scene demonstrates just how minimal the work undertaken to create the reserved track tramway was. The edge of the field fenced off, a bed of colliery ash on which sleepers were laid, with rails spiked to these. Earthworks, such as they were, were created to allow for any future track doubling.

Still looking east, but now just east of Wellsgreen, through which double track was laid. A cattle grid of triangular cut wood has been laid to prevent animals straying on to the tram lines, and a minor cutting allows an even track gradient.

Section feeder box (the door would normally be closed) with the private track on the south of the road between East Wemyss and Buckhaven, just east of Wellsgreen. These long single track sections (the next passing place was half a mile distant at Muiredge crossing) led to difficulty in working to timetable until electric signals were installed.

Above: The view west along the new tramway from Muiredge mineral railway crossing behind the camera, with the former toll house on the right at the end of the road to Perceval. The rail layout was designed so that the car going straight ahead normally had a direct run into the loop.

Below: Another of the former Bowman's pits was at Muiredge. The car is just starting along the first part of the new wide 'high road' constructed through undeveloped arable land. By providing this new highway – Wellesley Road – Mr Wemyss was enabled to close the old road along the shore then use that space for rail sidings and the colliery waste tip.

Looking in a westerly direction along the newly built Wellesley Road, with not a single building in sight; this very soon changed. On the left is a signal for the Wemyss & Buckhaven Railway's Methil line. The area on the shore became the site of the Wellesley Colliery.

From near the same spot but looking east with the newly-erected Baum Washer of the Wemyss Coal Company prominent on the right. When built it was said to be the largest such in Europe and probably in the world. Coal from most of the Company's pits was brought here for washing prior to export or sale.

One of the first buildings built on the new road was the White Swan Hotel, again to the design of A S Tod. Behind the tram is the parapet of the bridge which carried the new highway over the Wemyss Coal Company's mineral railway. From here the White Swan Brae descended, joining the old low coastal road at West High Street Buckhaven.

Looking east from Methil Cross Roads before Wellesley Road is nothing more than a sixty foot wide swathe across open fields. The then agricultural nature of the area is apparent, but in the distance can be seen the silhouette of Leven Colliery No. 1 and No. 2 Pits.

Above: The basic car depot at Aberhill, between the farm and Leven Colliery. Car number 4 has two tower wagons for company. One is rail mounted and owned by the tramway company, the other on the right is the property of Bruce Peebles. No attempt has yet been made to pave the roadway, although the line has been in use for a couple of weeks.

Below: Interior of the depot on the same day with cars 1 and 5. The latter is still to be fully fitted up, the trolley not yet fitted on its spring base. The pit in the foreground contains a trolley jack, used to lower traction motors from the trucks.

The very basic Aberhill Depot, recorded at the end of September 1906, taken from the John Patrick album. Built without any workshop facilities whatsoever, these were later added to the back end of lye number 1 on the right.

The major earthwork required for the tramway was the cutting at Kinnarchie Braes to bring the line down from the level of the raised beach which it had followed for much of the length of Wellesley Road down to the public road some 50 feet below. Whilst not a public thoroughfare, it soon became a handy shortcut.

Car 2 heading west on the double track round Leven's Branch Street. As seen from the previous view, the actual 'head' of the harbour – originally behind the wall on the left, which disapeared after 1888 when ownership of the dock passed to the North British Railway. They filled most of it with colliery waste to ensure coal traffic was shipped from their Methil facility.

With the camera turned in a more easterly direction but still at the same locus, the presence of the new tram on Branch Street is a source of fascination to some, but of total indifference to others. The former harbour area on the right is today a car park; the motor car- as almost everywhere – omnipresent.

Patrick concentrated much of his attention on the Leven High Street area and proceeded to make commercial postcards from several of these views. They form a fascinating record of the town's main thoroughfare over a hundred years ago.

Moving further up the High Street to the end of the double track passing place at Bank Street, on the left. The photographer has caught the everyday scene of the shop assistant at Gourlay's stationers on the left in the act of cleaning the windows. This was one of the town's oldest buildings, with the door lintel dated 1681.

Car 8 on the High Street at Forth Street corner making for the terminus. Behind the tram is the Caledonian Hotel, centre of local social life and scene of annual works' soirées – including that held by the tramway company for its employees. The tram track is, perhaps dangerously, close to the footpath.

High Street at Waggon Road, the name an indication this had been the site of the crossing of an early coal waggonway which ran from coal pits at Sillerhole to a salt works near the shore and the harbour. A busier scene with no less than three horse-drawn carts, but no impediment to treating the street as a playground.

Car 5 at the top of Durie Street, about to enter the final passing place before the terminus. The accumulation of local children assured that a good number of these cards would be purchased, a ploy frequently resorted to by postcard photographers.

The opening of the Wemyss tram line coincided with a surge in popularity of the picture postcard and publishers were not slow to cash in on the new tram phenomenon. Eight cards for 4d was, even then, quite a bargain.

THE NEWEST SERIES
—OF—
LEVEN POST CARDS
—SHOWING—
The TRAMWAYS
At 8 Different Points from Leven Bridge to Durie Street.

SET OF EIGHT DIFFERENT VIEWS
—FOR—
FOURPENCE.

SPECIALLY PRODUCED FOR
MALCOLM'S
STATIONERY SALON,
80 HIGH ST., LEVEN.

In addition to the eight views from Mr Malcolm, there were several 'Glimpses along the route' postcards, plus this multi image summary. All views featured were available as individual postcards, and it is probable that eight were sold as a 'set', published in their 'Ellenslie' series by T G Blyth of Kirkcaldy.

The track of the Wemyss line commenced by an end-on junction with the terminus of the Kirkcaldy tramways in Rosslyn Street, Gallatown. The Wemyss line turned right into Randolph Road in front of the two storey house on the right side. The printer has produced a very reasonable representation of the colour of the tram.

Double track took the line round into Randolph Road, where it was laid (for the most part) on the 'waste' on the north side. Several attempts were made to purchase a strip of land in the field to allow construction of a reserved track. The proposed line to Thornton and beyond was to keep to the main road to the left here.

The first settlement along the line deemed worthy of the postcard publisher's attention was Coaltown of Wemyss; again the local children are assembled for potential sales generation. This scene at the present entrance to the Wemyss Castle policies is where the famous Wemyss School of Needlework is situated, founded by Dora Wemyss in 1877.

Moving east along the line, but still in Coaltown, these dwellings were probably constructed for use, not by miners, but estate workers. Prior to the coming of the tramway, most of the village lay south of Main Street, but subsequent development was to the north. In recent years considerable effort has been invested in revitalising the old dwellings.

Next village arrived at was East Wemyss, with the barefoot children arrayed in front of tram car number 3. Again at this time the village had developed between the high road and the shore, to the south. Many of the houses were occupied by miners employed at Michael Colliery; opened in 1897 it grew into one of the largest and most productive in Scotland.

East Wemyss looking west from the top of East Brae. Most of the old structures beside the Kingslaw Burn which passes below are long since removed. This small settlement grew along the High Road which was created in the 18th century as an alternative to the tortuous coastal route via the East and West Braes.

Evocative card recording a scene of which nothing remains. On the right the buildings of the Rosie Colliery, on the left Wellsgreen Terrace housing built for its workers. As road traffic increased this corner became a dangerous hazard.

The construction (on the left) of the brand new two mile highway through the fields for the tramway allowed closure of the original coastal route. The nearest Buckhaven dwellings can be seen in the distance; the old road is on the extreme right, with alongside a later road dipping to pass under the rail bridge at the station.

Wellesley Road was named to recognise the maiden name of Mr Wemyss' second wife, and also the name given to the redevelopment of the colliery, the buildings of which can be seen appearing on the right. The first workers' houses of the entirely new village of Denbeath were mostly inhabited by miners from the west of Scotland.

The then western boundary of construction of this new village in 1906 was here at School Street (now Wall Street) and the site chosen for construction (three years later) for the hospital named in memory of Randolph Wemyss. As with most local buildings of that era it was designed by the prolific estate architect A Stewart Tod.

The corner of Wellesley Road in Denbeath overlooked the teeming industrial scene on the shore below with busy pit and constant movement on the railway sidings. Developed into one of the largest worked by the Coal Company, it closed in July 1967. The site, having been used for several years subsequently for construction of oil platforms, is now fabricating wind turbines, all progressions of the energy industry.

From Kinnarchie Braes the panorama over the River Leven to the town and beyond demanded the attention of photographers. The result was that, counting only views with a passing tramcar, there are at least a dozen variations. The former toll bridge (the toll was a 'Bawbee', or half-penny) was built in 1840, the toll removed 30 years later and the structure itself replaced in 1957.

The east end of Leven High Street with a remarkable candy-striped brick gable of the building on the left, which now houses Leven's Library. At this time the High Street had an eclectic mix of buildings, some dating back to the 17th century, plus an assortment of Victorian additions

The end of the line was here in Durie Street at the gates to Carberry House. There were several proposals for an extension as far as Scoonie, but these never materialised. Carberry was built for the Balfour family in the late 19th century and given by them to the town in 1929.

![Photo of car 3]

Photo of car 3 from the John Patrick album. It seems likely that this was prepared for Stephen Sellon, the engineer for the line, as several of its scenes were used in contemporary technical journals, with acknowledgement given to him.

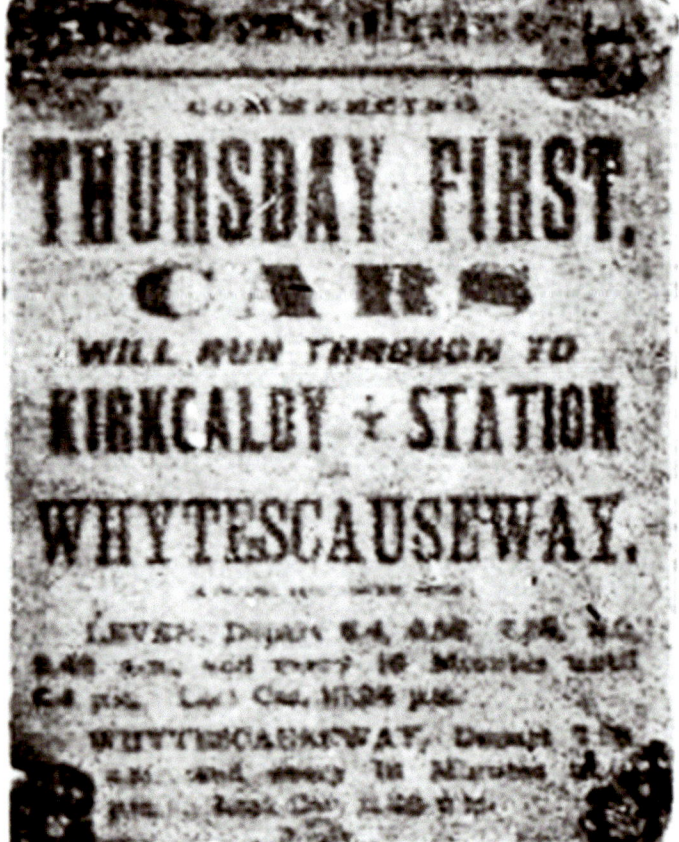

It is possible to date the above photograph by enlarging the poster in the car window which refers to operation to Whytescauseway 'commencing Thursday first". This happened on Thursday 27th September 1906, achieving one of the fundamental aims of the promoter, assisting communication with Kirkcaldy railway station.

WEMYSS AND DISTRICT TRAMWAYS.

SUNDAY.

Leven (depart), 7.20, 8.32, 9.20, and 9.44 a.m., and every 12 minutes till 7.56, 8.20, 8.44, and 9.8 p.m. Last Car, 9 20 p.m.

Whyte's Causeway (depart), 8.20, 9.32, and 10.8 a.m., and every 12 minutes till 8.20, 8.44, 8.56, and 9.20 p.m. Last Car, 10 20 p.m.

MONDAY to FRIDAY (inclusive).

Leven to Whyte's Causeway, 6 a.m., 8 a.m., 9 a.m., and every 30 minutes until 10 p.m.

Leven to Gallatown, 6 a.m., 8 a.m., 9 a.m., 9.30 a.m., 10 a.m., and every 15 minutes until 9 p.m., 9.30 p.m., 9.45 p.m. Last Car, 10 p.m.

Whyte's Causeway to Leven, 7 a.m., 8 a.m., and every 30 minutes until 9 p.m. Last Car, 11 p.m.

Gallatown to Leven, 7.20, 8.20, 8.50, 9.20, 9.50, 10.20, 10.50, 11.5 a.m., and every 15 minutes until 9.35 p.m. Last Cars, 10 35 and 11.20 p.m.

EXPLANATORY NOTE.—Cars from Leven leaving at the hour and half-hour run through to Whyte's Causeway; those leaving at the quarter after and quarter before the hour run to Gallatown only.

SATURDAY.

Leven (depart), 6.8, 6.32, 6.56, 7.20, and 7.44 a.m., and every 12 minutes till last Car, 9.56 p.m.

Whyte's Causeway (depart), 7.8, 7.32, and 7.56 a.m., and every 12 minutes till 10.20 and 10.44 p.m. Last Car, 10.56 p.m.

TIME-TABLE OF WORKPEOPLE'S CARS.

Leven (depart), 7.10, 8.30 a.m.; Shorehead, 4.54, 5.18, 7.14, 8.34; Aberhill, 4.50, 5.0, 5.54, 7.20, 8.40; Denbeath, 4.54, 5.4, 5.24, 5.58, 7.24, 8 44; Buckhaven, 4.56, 5.6, 5.26, 6.0, 7.26, 8.46; Rosie, 5.1, 5.11, 5.31, 6.5, 7.31, 8.51; East Wemyss, 5.7, 5.17, 5.37, 6.15, 7.37, 8.56; Coaltown, 5.22, 5.44, 6.20, 7.42, 9.1; Gallatown (arrive), 5.30, 5.50, 6 28, 7.49, 9.9.

Gallatown (depart), 5.10, 5.32, 6.0, 6.30, 7.50, 9.18 a.m.; Coaltown, 5.18, 5.40, 6.13, 6.38, 7.58, 9.26; East Wemyss, 5.9, 5.23, 5.45, 6 21, 6.43, 8.3, 9.31; Rosie, 5.17, 5.31, 5.51, 6.27, 6.49, 8.9, 9.37; Buckhaven, 5.22, 5.36, 5.56, 6.32, 6.54, 8.14, 9.42; Denbeath, 5.23, 5.38, 5.58, 6.34, 6.56, 8 16, 9 44; Aberhill, 6.2, 6.38, 7.0, 8.20, 9.48; Shorehead 5.47, 7.6, 8.26, 9.52; Leven (arrive), 7.9, 8.29, 9.58.

Cars are run at intervals during the day between Leven and Gallatown to meet requirements of the different colliery shifts.

Timetable dated 8th November 1906, giving times of cars running through to Whytescauseway in the centre of Kirkcaldy. Details of workers' cars are given, the first leaving the depot at 4.50 am. That scheduled to leave Gallatown at 5.10 am was kept overnight in Gallatown Depot.

To cope with the unanticipated volume of passengers, the need for additional cars was debated as soon as 7th September 1906 (before the official opening). They were ordered by early November and delivered the following March. With undefined 'improvements', the only apparent change being no offside step for access to the platform, this is car 11 when new.

The four cars bought by the Wemyss Coal Company for use by their workers were considerably longer than the 'mustard boxes' and had eight removable side panels, seen in this view. Each panel had a lifting handle near the bottom edge and the four central compartments had steps to assist access. They were sold to the tram company at the end of 1912.

Wemyss car 13 is standing on the single track terminus at the foot of Whytescauseway in the early days of through operation. From here the initial timetable allowed 64 minutes for the near ten mile journey to Leven. This was found impossible to achieve in normal service and was soon increased to 75 minutes. After much of the Wemyss track was doubled this was reduced to one hour in 1914, with a through fare of 6d.

The foot of Whytescauseway was where Kirkcaldy's 'Upper Route' met the High Street line. The through running agreement restricted Wemyss cars to this part of the town's system; Corporation car 9 is at the terminus. The large linen works building behind the tram was offered to the Wemyss undertaking for a potential depot site.

At the top of Whytescauseway, Wemyss car 8 will turn right in to Wemyssfield. The 'baronial' building on the left is Kirkcaldy's original Sheriff Court House.

At the eastern end of the Upper Route was a single track triangular junction connection with the original Kirkcaldy route here at Junction Road. This must be in the earliest days of through operation as the Wemyss car has its original trolley fitting. This had to be replaced to raise the base of the trolley to the same height as that of the Kirkcaldy cars.

The mile length of Kirkcaldy's track from Junction Road to Gallatown was single, with four passing places. One of these loops is behind the car in this view of St. Clair Street. Despite fitting of signals, the single track caused operational difficulties, but was never improved.

As deliberate policy to save money, the Wemyss line when built had no workshop facilities. To provide stabling for early morning workmen's cars, one or two Wemyss cars were kept overnight at Gallatown Depot, at a cost of one shilling per night. This scene shows a Wemyss car at Kirkcaldy's Gallatown depot.

When the joint through running operation started, and for some weeks prior, the severe shortage of cars on the Wemyss' line required the hire of up to four Kirkcaldy cars. This is Corporation 14 at the impressive tram entrance to the Wemyss estate at North Lodge. The line ran through fields to Bowhouse Farm, cutting off a long detour required by road.

The Leven panorama from Kinnarchie Braes, this time with a Kirkcaldy tram in the scene. The density of the original part of the town is very apparent. Now unrecognisable; the old harbour area is transformed with new roads, the bus station and the leisure centre .

The top end of Leven High Street; to judge by the high level of interest which Kirkcaldy car 21 is exercising, such a sight was then still a novelty. It is open to debate as to whether this is a car hired to the Wemyss company, or one of the first Kirkcaldy cars on the joint through running service.

Kirkcaldy Corporation number 16 near the terminus in Leven. The date of this scene (1906) suggests that this is one of the cars hired to overcome the shortage initially experienced as a consequence of unanticipated demand.

THE WEMYSS AND DISTRICT TRAMWAYS COMPANY, LIMITED.

PARCELS SERVICE

WILL commence running on 1st May 1907. Agents have been appointed to receive, collect, and deliver Parcels in the following Districts:—

District.	Depots and Agents.
LEVEN,	CALEDONIAN HOTEL COACH OFFICE—*Telephone, No. 0100.*
ABERHILL,	TRAMWAYS OFFICE.
METHIL, INNERLEVEN, and DISTRICT,	G. MARR, Carting Contractor, Methil.
BUCKHAVEN, for Buckhaven, Links, Denbeath, and Muiredge,	THOS. BONTHRON, Stationer, Randolph Street, Buckhaven.
EAST WEMYSS, for East Wemyss and Rosie,	R. SIMPSON, Grocer, East Wemyss.
COALTOWN OF WEMYSS,	RUSSELL BROS., Grocers, Main Street, Coaltown.
BOWHOUSE TOLL, for West Wemyss,	G. DEWAR, Bowhouse Toll.
KIRKCALDY, for Kirkcaldy, Sinclairtown, Gallatown and Dysart,	Messrs J. ROSS & Co., Contractors and Carting Agents, 52 Townsend Place, Kirkcaldy—*Telephone, No. 127.*

SCALE OF CHARGES (PREPAID ONLY)—

Not exceeding 14 lbs. 28 lbs. 42 lbs. 56 lbs
3d 4d 5d 6d and 1d for every additional 7 lbs.

LIABILITY:—Up to £2 only, unless declared, and extra charge paid for Insurance.

All Goods will be carried under the conditions as set forth in the Carriers Act, 1830.

At present a PARCELS SERVICE will run as under:—

Leave LEVEN, 8.30 a.m. Leave GALLATOWN, 9.30 a.m.
,,   2.0 p.m.   ,,   3.0 p.m.
,,   5.50 p.m.   ,,   6.50 p.m.

Express Delivery will be made, and a Parcels Service will be run from time to time to meet the requirements of the Public.

Further Particulars can be obtained on application to any of the Company's Agencies.

W. T. J. MARSHALL, General Manager.

TRAMWAYS OFFICE, ABERHILL.

N.B.—When you have Parcels for despatch, notify the Agent of your District.

An early feature of the new tramway was its parcels delivery service. For this purpose two small vans were purchased. From circumstantial evidence it had been deduced that these originated with the Potteries tramways, where Stephen Sellon was also the engineer.

The Potteries tramway purchased six of these small trailers; the total in official returns reduces to four in 1907.

The following Accounts were passed for payment out of Working Capital Account, viz:—
1. *Potteries Electric Traction Co. Ltd. for two baggage vans* £142 : - : -
2. *North British Railway Co. for carriage of same* 8 : 4 : 10

WEMYSS AND DISTRICT TRAMWAYS COMPANY LIMITED.

HOLIDAY NOTICE.

THE undermentioned SERVICE of WORK-PEOPLE'S CARS will NOT RUN from July 23rd to July 27th (both days inclusive):—

LEVEN Depart.	ABERHILL Depart.	DENBEATH Depart.	GALLATOWN Depart.
	4.30 a.m.		5.10 a.m.
	4.45 ,,		5.25 ,,
	4.50 ,,		5.35 ,,
4.54 a.m.	4.57 ,,		6.5 ,,
		5.22 a.m.	
	4.52 p.m.		5.30 p.m.

The PARCELS SERVICE will be entirely Withdrawn on July 23rd and 24th.

THURSDAY, 23rd July, being Market Day at Leven, Cars will Run to SHOREHEAD only

W. T. J MARSHALL,
General Manager.

Aberhill, July 21st, 1908.

Top: Confirmation of the sale was found in an old Wemyss tramway ledger. An entry of 16th April 1907 records payment of £142 for two baggage vans and £8.4s.10d to the railway company for their transport.

Above: In Fife the vans performed useful service until the end of November 1914 thereafter languishing beside the depot. A home was found for one, as a garden shed, behind a house on Wellesley Road.

Major Marshall made good use of local newspapers to advise changes. The local holiday period (the last week of July) brought suspension of the workers' early morning services as detailed.

Kirkcaldy Corporation trams were extended to serve Dysart from the end of January 1911. Parliamentary powers existed for this line to continue up Normand Road to the railway level crossing for Francis Colliery. It was hoped that the Wemyss line would be extended down Boreland Road to join this, but the Company saw no benefit.

Car number 1 makes the turn into Randolph Road at Gallatown. The extension to meet the West of Fife tramways would have continued north here on the main road, and then turned west in Thornton along Strathore Road.

Development of the Aberhill area followed rapidly. The building occupied by the Clock Tower Tea Room and Public House was built at the laird's expense and presented by him to the tram company during 1907. Aberhill Farm buildings, on the left, were adapted as offices, with a new gable built parallel to the new road.

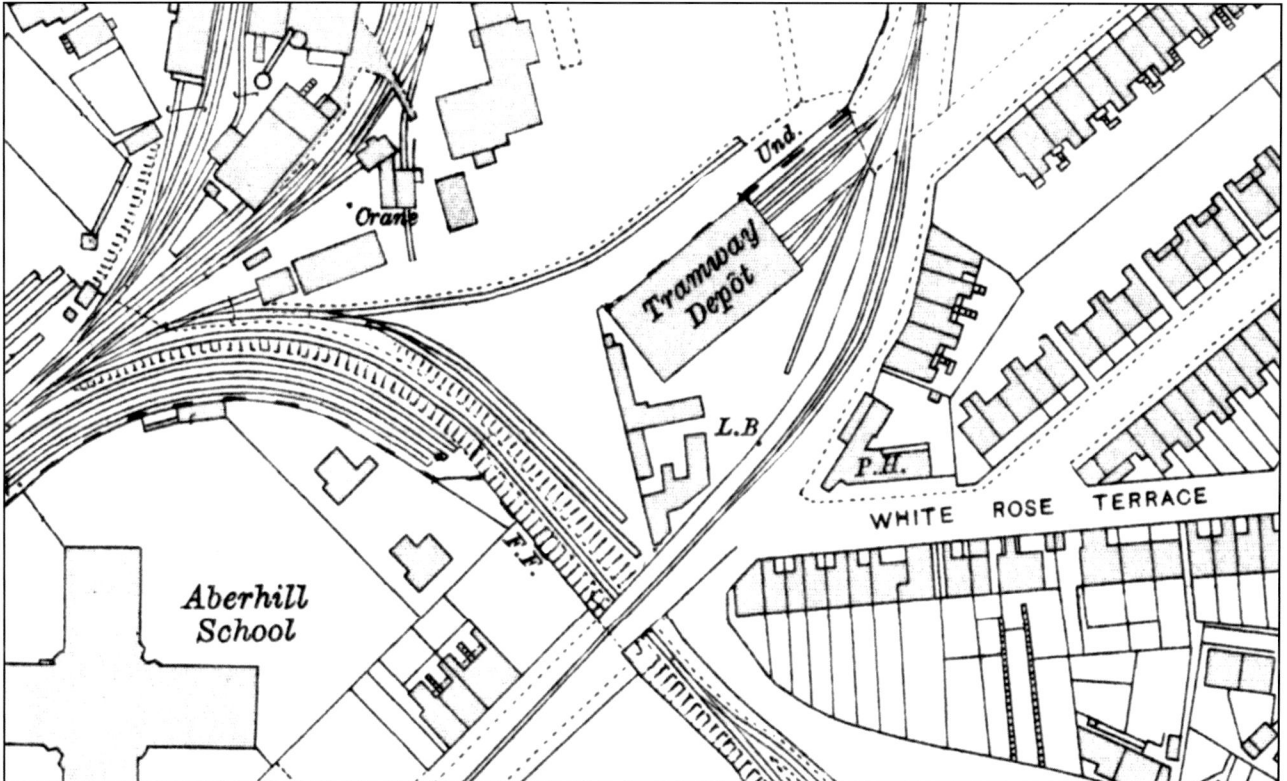

Plan of the depot and surrounding area. The way that the corners of the Aberhill Farm buildings and the first house on White Rose Terrace have been 'trimmed' to accommodate the new line of the road is apparent. The area south of the depot became the site for the 1920s bus garage.

The track doublings which improved timekeeping and reduced journey times involved considerable planning. Here in Wellesley Road, looking west from Cowley Street, the new construction on sleepers was more able to accommodate mining subsidence.

When the programme of track improvement was completed schedules were reduced by 15 minutes. This looks east from the same point above and shows the narrow road bridge over the mineral railway which was widened at this time.

The most significant engineering work on the tramway was the curved cutting down the hillside from Aberhill down to the bridge over the River Leven. This was always double track.

The end of the reserved track in the cutting where it became single track to cross the narrow bridge. As other road traffic increased this became an ever-increasing hazard.

A car passing the entrance to Wemyss Castle Station (on the right). This was one of the earliest sections of track to be doubled, as far as the 'Car Shed'. The road on the right led up to the station where a private waiting room was maintained for use by the laird and his guests .

The remarkable transformation of Wellesley Road seen in 1910 – a mere four years since this was open agricultural land. As is apparent, from the outset the street was well provided with shopping facilities. The distinctive (but untypical) block on the left between Bayview Crescent and Kirkland Road is recognisable today.

Many of the streets used by the trams negotiating Leven were narrow – part of the High Street measured just 17 ft wide. Originally in Bridge Street the line was single. However, operational problems were eased in 1914 following installation of a second track (the outer one round the curve)

A view at the terminus about the same date. Car 3 is now just part of the scenery, an unremarked component of everyday life. Keeping a close eye on the photographer (from the large postcard publishers, Valentine of Dundee) the kilted scholar on the right is probably homeward bound, the shadows indicating afternoon rather than morning.

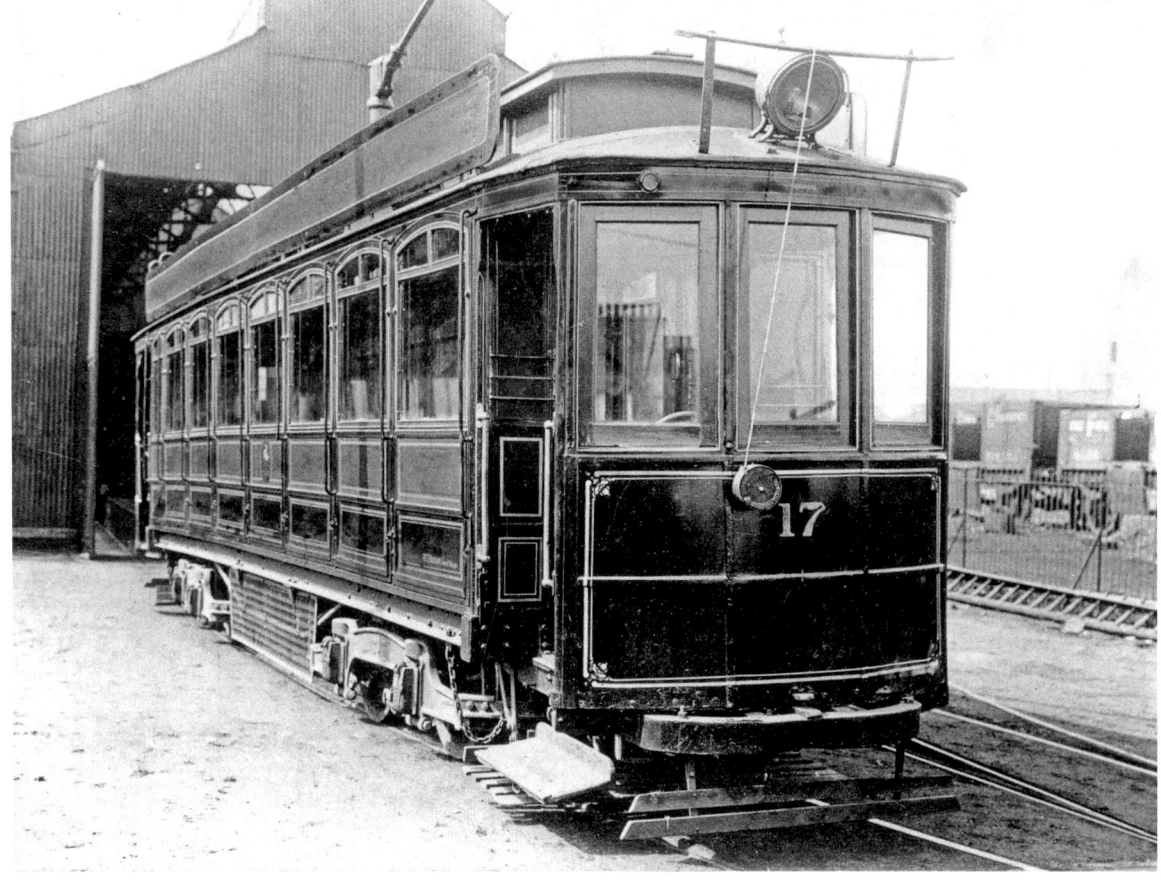

Car 1 at Aberhill after change of livery from Wemyss yellow to Balfour Beatty mahogany lake in April 1913. The way that early photographic emulsion reacted to the different colours is shown by the gold lining standing out on the lake colour, but almost invisible on the yellow. (Compare with the photo of car 4 on page 33.)

The long workers' cars underwent several alterations. The removable side panels were fixed permanently in position with the lower handles removed although the vertical hand holds remained. Track brakes were removed from the bogies and side steps removed. The livery change emphasises the lining details.

By 1914 track along Wellesley Road track had been doubled. While many of the original structures on the south side of the road have since been demolished, these distinctive Denbeath buildings remain. The car (number 17) had then recently been transferred from ownership of the Coal Company to the tramway undertaking.

Aberhill rapidly became a busy transport hub, the shelter, seen on the right, greatly appreciated by transport staff changing shifts. Workers' car 15 is in virtually original state, still with its steps but having lost its roof-end advert boards.

At the opening of the tram line, platform staff had uniforms of navy serge with piping in mustard yellow; the circular cap badge conspicuous in photographs. The new ownership in 1912 brought change, uniforms as shown here, with brass initials W&DT on the lapel; cap badge no longer worn.

Mr Dawson surrounded by his platform staff outside Aberhill Depot, probably in the early 1920s. In addition to four Inspectors there are 18 motormen and 21 conductresses (but not one male conductor). Chief Inspector James Mitchell, who served the tramway from first day to last, is to Mr Dawson's right.

Share certificate issued to the manager (Mr Dawson) on 27th September 1921 for one hundred 7½% Preference Shares. One can only hope that he knew if this was a good investment.

When the Wemyss tramway opened, in the name of economy, it was decided to have no repair workshop, but to make use of the Corporation's facilities. This was found to be a false economy, so when Aberhill depot was extended, a small workshop was built at the end of lye 1, seen here. £220 was spent on machine tools.

Two state of the art trams were added to the fleet at the start of January 1925. From the Brush Company of Loughborough, they had been mooted as early as 1920 as long single-truck cars, but the design evolved into what is seen here.

In January 1932 both of these cars were transferred to the associated Dunfermline and District Tramway Company, another Balfour Beatty subsidiary. They were little used there, and on closure in 1937, they were kept for potential, unfulfilled, resale.

Housing development followed the tram lines and these attractive detached houses were built at the western end of Wellesley Road in the 1920s.

A rude awakening for tram driver Pete Ferrie and his conductress Beth Finnigan (the only persons aboard) on the last car on 25th May 1926. An obstruction on the rail launched the car into the Den at East Wemyss and only a massive tree prevented a forty feet fall. This was a time of industrial unrest, but there was no suggestion that this was a result of any deliberate act.

An early aerial record of Muiredge Pit. On the right hand side the line of the tramway passes (as single track) in front of the "Buck and Hynd" public house, cuts the corner on reserved track then heads across to the start of Wellesley Road , top right, before construction of council housing.

A scene full of fascinating detail. Top left is Aberhill tram depot and bus garage. Central is Bayview Park, home for many years to East Fife Football Club. A game appears about to commence, with the supporters present but no teams on the pitch. A surprising number of supporters have come by car, parking in the surrounding streets. .

Possibly recorded on the same flight; the panorama of Leven in tramway days stretches beneath. One area in particular which has seen massive change is around the harbour; which by then had been totally infilled, while the tramway down from Kinnarchie Brae is now incorporated into a new road.

The town's gas holder then dominated this northward vista from Kinnarchie Braes, across the River Leven to Largo Law on the far horizon. The tramway interest is provided by one of the large workers' cars crossing the 'Bawbee Brig' en route to Gallatown.

Evolution of the 'Workers Cars': *Above*, Car 16 still in unaltered state in August 1910 (date courtesy of the advert for 'The Spadoni Sisters' at the Gaiety Theatre). *Centre*, by 1920 the initial changes had been made, side panels no longer removable, brake rigging simplified and side steps removed. *Lower*, Final condition with sides now panelled, headlamp lowered and new destination box.

The first three buses ordered by the Tramway Company were of this type; Tilling-Stevens with 29 seat bodies built by Strachan & Brown in London. SP7216 was the first. Comfort of passengers was increased immeasurably after fitting with pneumatic tyres.

The Wemyss & District Tramways Co., Ltd.

OMNIBUS SERVICE.

LARGO & KIRKCALDY (GALLATOWN)
via. LEVEN, BUCKHAVEN, &c.

DAILY SERVICE

commencing on

FRIDAY, 21st September, 1923

See Bills for further particulars.

W. T. DAWSON,
General Manager.

The Tramway Company rose to the challenge of motor vehicle competition, with their first buses purchased in July 1922. Compared to the life of a tram often stretching to several decades, these first vehicles gave only six years service.

Tilling-Stevens SP8086 waiting in College Street Buckhaven outside the original High School. Taken at Church Street cross roads, with one of the Burgh's constables on hand to quell any scholarly over-enthusiasm.

Mr Dawson was keen to make proper record of his staff, included cleaning and maintenance employees equally. An annual soirée was held for staff and guests, where he was always lavish in praise for his "happy and hospitable" employees.

After the initial six Tilling-Steven buses came three charabancs, typified in this 1924 view of SP9168 with the Wellesley Colliery brass band. The prize-winning musicians travelled considerable distances to perform, being particularly popular at galas at South Shields, Tynemouth, and mining communities in the north east of England.

Unfortunately the Tilling-Stevens was found to be rather ponderous to compete with smaller, lighter vehicles of competitors, so to compete on more equal terms Mr Dawson was authorised to purchase four of these handsome Halley 20-seat buses, manufactured in Yoker, Glasgow. The drivers knew them as the *'Chasers'*.

Photographs of the 'competition' vehicles seem more elusive. This is FG989, an American 'Reo' Pullman 24-seat bus of 1925, one of five similar owned by Thomas Anderson and James Roden trading (then) as A & R Pullman Motor Safety Company. A thorn in the flesh to the Tramway Company, they were bought over in February 1926.

Press advert relating to through bookings from Anstruther to Glasgow. What is not stated is just how long such a journey took.

The next bus order was placed with Albion – another Glasgow manufacturer. FG1939 is from a purchase of six vehicles with 25-seat bodies. This was a favourite place to pose – in front of the depot/garage at Aberhill.

Sylvan scene c.1928 with one of these Albions on the route to Leven from St. Andrews, posed here in Kingsbarns by the attractive surroundings of the 17th century Parish Church.

Last day of operation of the Potteries Tramways. It was said that car 82 seen here became Wemyss 20, but it is apparent that the position of the clerestory ventilators does not correspond with those on that Wemyss car seen below. Major changes made at Aberhill are apparent, particularly addition of drivers' vestibules and provision of new trucks.

In May 1929 two tram bodies were purchased, formerly operated by the Potteries Electric Tramways. After considerable alteration they became Wemyss numbers 20 and 21. This view of the former car is at Leven terminus, Carberry gates. In the final months of running most operation was by bogie cars 14 – 21.

The National Steel Foundry was responsible for casting of many components for the two English cars in 1929. Completely new equal wheel bogies were made, based on the 1907 Mountain & Gibson units under the workers' cars. When the two 1925 built cars went to Dunfermline in 1932, they were equipped with these trucks.

The two cars from the Potteries line, upgraded and fitted with drivers vestibules fitted very well into the Wemyss fleet and had a distinct family resemblance to the older cars. This enhanced scene gives, with the benefits of current technology, a representation of the livery, using colour samples retrieved from car 16 at Falkland.

The same colouring process applied to a scene at Leven terminus recorded by Dr H A Whitcombe in July 1929. Cars 8 (behind) and 9 show variations, one with its rocker panels and window frames off-white.

Wemyss number 16 is seen in May 1931 at the Gallatown terminus, the Turret House Tavern. The appearance of these cars has altered, the removable panels dispensed with and the sides divided with waist and rocker panels, no track brakes, the light repositioned to the centre of the dash and destination boxes fitted

These three photographs at Gallatown were taken to record the demise of the Kirkcaldy Corporation trams on 15th May 1931. Car 22 dated from 1904 and ran unaltered for its entire twenty-seven year existence. No surprise then that Kirkcaldy chose to dispose of their trams rather than contemplate upgrading them.

Kirkcaldy owned four more youthful trams, built in 1915-16. These had the luxury of being fitted with drivers' vestibules to the platforms. However, they were unpopular with drivers as the screens created draughts down the stairs! It is probable that both these cars saw use on the Wemyss lines.

Four cars – of which this is one – were passed from Kirkcaldy to Wemyss and in use by early July 1931; these were Kirkcaldy 23-26. They had top deck seats and stairs removed and the new owner's name painted on the waist panel.

A second batch of four Kirkcaldy cars was then transferred, in service on the Wemyss lines from October 1931. These had considerable alteration before use; drivers' vestibules were made plus the alterations made to the earlier batch. Car 28 had a further working life of just four months.

Car 14 at speed on the 'semi' reserved track of Randolph Road on 24th January 1932, less than a week before closure. This car became, after the end of operations, a hen house at Blacketyside Farm between Leven and Lundin Links.

Number board retrieved from the inside end of the saloon clerestory. It is possible, with a high degree of imagination, to consider that the faint outline of the top part of a figure '3' could be made out between the two figures on the right. Examination with infra-red light could possibly resolve matters.

Rails were rapidly removed from streets. Here in Wellesley Road the concrete strips on both sides of the former track remain, but the rails have gone, except for a short length visible remaining to be dealt with. The bus is a Leyland Lion (WG282) of 1931, then licensed by the General Motor Carrying Company.

Aberhill Depot was used, little altered, for several decades as a bus garage. The Leyland Lion (FG9455) was new in 1934 to the General Motor Carrying Company of Kirkcaldy and gave over twenty years service.

The office buildings at Aberhill were adapted from Aberhill farm steading. Gables were rebuilt to follow the new road, giving an unusual 'sliced-off' appearance. Heavily disguised as an indoor bowling facility the depot buildings remain, but the offices have been demolished. Signs of mining subsidence are apparent. *Photographs: The late G N Heathcote*

Leyland WE7067 was used on local services in the 1930s. It started life with Sheffield Corporation and is pictured at the bottom of Leven's North Street, standing by the art deco gas showrooms. On the left is the gable end of the Star Hotel.
Allan Condie collection

While the Tower at Aberhill, and the bar, remain, the shelter has gone, also the row of dwellings to the right of the tower. The bus, CAV175, had a slightly unusual pedigree, set aside from the 'average', by having doors to the rear platform. It had been supplied new to Sutherland of Peterhead, then transferred to Aberhill garage in 1951-2.